D0207773

Concise Complex Analysis

Concise
Complex
Analysis

Sheng Gong

University of Science and Technology of China

W **World Scientific**

Singapore • New Jersey • London • Hong Kong

Published by

World Scientific Publishing Co. Pte. Ltd.
P O Box 128, Farrer Road, Singapore 912805
USA office: Suite 1B, 1060 Main Street, River Edge, NJ 07661
UK office: 57 Shelton Street, Covent Garden, London WC2H 9HE

British Library Cataloguing-in-Publication Data
A catalogue record for this book is available from the British Library.

CONCISE COMPLEX ANALYSIS

ISBN 981-02-4378-2

Printed in Singapore.

47105893

Preface

Professor Gong has written a very appealing book on complex analysis. It is indeed concise but this is far from its only attribute. ("Concise" applies to the discussion of those results in complex variables which are completely analogous to real-variable calculus results, and to Professor Gong's ability to extract the essence of a proof in his presentation.) The book is also insightful, and when an important result can be viewed from different angles, the author rightly feels that it is valuable to point this out. The author's viewpoint is not only that of a geometric function theorist but that of a very broadly trained and broadly published analyst.

There are three noteworthy features of the choice of subject material. First, Cauchy's theorem is treated from the point of view of Green's theorem as well as via the Goursat proof. As well as being pedagogically sound, this permits an introduction to d-bar techniques which are used for example in proving the Mittag-Leffler theorem. Second, there is a beautiful chapter entitled "Differential Geometry and Picard's Theorem", which contains a discussion of Gaussian curvature of conformal metrics, the Ahlfors-Schwarz lemma, proofs of Liouville's theorem and Picard's first theorem via the explicit construction of conformal metrics, a geometric disucussion of normal families, and Picard's second theorem. Third, there is an introduction to several complex variables which illustrates some of the differences between the one-variable theory and the several-variable theory.

There is more than enough material for a one-semester course, in fact in just 170 pages the reader is taken from the basics to topics which are often part of a second course in complex analysis. There are a great many excellent exercises.

Professor Gong is the student of L.K. Hua and a former Vice-President of the University of Science and Technology of China. He is the author of many books, a number of which have already been translated into English. The English edition of the present text will be a valuable addition to the advanced undergraduate and beginning graduate textbook literature.

Ian Graham
University of Toronto
May, 2000

Foreword

I have been teaching for years theory of functions of complex variable both at home or abroad and therefore considering compiling a textbook for the course. Known as a subject of a long history, there have appeared numerous textbooks on the theory of functions of complex variable, and there has been no lack of excellent ones. Then, why should I have worked out a new one? What was the guiding principle I had in my mind when I got the job started? What are the differences between the traditional textbooks and mine? The compilation of a new textbook, without well clarified principle of the author's own, would have probably turned out to be a case of parroting others' ideas or a matter of knocking different pieces together or borrowing the views from other textbooks of the kind.

It is a general trend to modernize mathematical teaching material today. But how to realization modernization awaits continual attempts and explorations. Only by doing this might we have the chance to see the correct approach emerge. However, we are reluctant to find a kind of somewhat doubtful practice of consistent additions of the modern mathematical contents to the elementary course. Thicker and thicker textbooks do not in any way imply that it is the right way to follow. Ultimately, basic course textbooks should be focused on the fundamentals. Therefore, the present book aims at recounting and testifying some of the traditional and elementary content in terms of modern concepts and language. With the rapid development of mathematics, however, some relatively modern ideas and theories are to be categorized as basics, the connotation of which are always changing with the time and the tide. My attempts in doing so, as I can foresee, will bring about disputes of various kinds. It doesn't matter. For we know that progress and development are usually made through controversies and disagreements.

An increasing number of scholars in the mathematical world have come to emphasize the unity in mathematics, i.e., they emphasize the interactions and interpenetration among different branches in mathematics. As a matter of fact, mathematics itself is unified and there is something of each in the other. The time is gone forever when the basic courses of different branches were severed and clearly cut from each other simply for the so-called pureness.

Within less than 200 pages, most content of the present book still conforms to the traditional approach. Meanwhile, the rest of the book is given to the elaboration of the ideas expressed above. It is well demonstrated in the following explanation on the disparities between this book and the traditional ones.

1. Complex analysis refers to the analysis in the field of complex numbers, to be more exact, it is the analysis in the complex manifold. Due to the historical reasons, complex analysis has often been termed as theory of functions of complex variable or analytic function theory. In recent publications, it is more often called complex analysis which is a more precise term. In this book, more than functions are meant when algebra, geometry, etc. are emphatically introduced into the content.

What should be the Complex Analysis in the college textbooks? It is the discussion of calculus in the complex number field, which is the first point of view of the author on this book. According to this point, the content of the college complex analysis is to be divided into two parts: one can be derived, without great effort, from the corollary of the calculus in the field of real number, i.e., the general college calculus. The other doesn't exist in the calculus in the field of the real numbers and cannot be obtained directly from the generalization. The former certainly counts while the latter is even more important. The first chapter of the book deals with the first part, the results of which can be, as has been mentioned, obtained directly without much difficulty. Therefore, some of the results are just plainly stated, with their proof processes left out. The content of calculus is abundant. It seems impossible or unnecessary to have them verified one by one here. Whether they can be extended to the field of complex numbers is decided by our firm grasp of their pivotal part. Then, what is the main part of the calculus? I have clearly pointed out, in another book of mine, "Concise Calculus", that calculus consists of three parts, i.e., differential calculus, integral calculus and the realization of the fact that differential and integral are a pair of contradictions. The third point can not be successfully elaborated until the application of exterior differential form, which was expounded in my recent writing "Making calculus teaching easy". In the light of the point, in the first chapter of the book, we explain the generalization of the results of the discussion in complex number field in addition to a very brief review of calculus. In this chapter, the fundamental theorem in complex plane turns out to be the complex form of Green theorem (§1.4 Theorem 2), which lays the foundation for the establishment of the Pompeiu Formula in the second chapter.

2. Traditional theory of functions of one complex variable comprises three parts: Cauchy theory of the integral, Weierstrass theory of series, geometry theory of Riemann, all of which, peculiar only to the complex number field, are absent in the field of real numbers or in the college calculus. The present book, as a college textbook should therefore reasonably include the three parts mentioned above, which in turn respectively make up Chapters 2-4 of the book. These chapters cover more than described in the Teaching Syllabus for Undergraduates. As has been elucidated, the book differs from the traditional ones in that part of the traditional content has been modernized this way or that way in this textbook.

From the point that theory of functions of one complex variable is, in the essence, the calculus in the complex number field, the fundamental theorem of calculus proves to be the Green theorem of complex form in the complex number field, which promptly leads to Cauchy-Green Formula, i.e., Pompeiu Theorem (§2.1, Theorem 1, Chapter 2). Now the real part and imaginary part of the function belong to C^1, unnecessarily asking for that the function should be holomorphic. Here the Cauchy's integral theorem and formula naturally become its result by simple inference. Then, why should the Pompeiu Theorem be first introduced instead of the Cauchy's integral theorem and formula? It is because: 1. It is just to follow a logic train as a matter of course from the point that theory of functions of one complex variabe is the calculus in the complex number field. That is, from complex form of the Green theorem, the result of the inference, unconditionally, should be Pompeiu Formula, not the Cauchy's integral Formula; and 2. The solution of one dimensional $\bar{\partial}$ equation (§2.1, Theorem 4, Chapter 2) could be inferred by Pompeiu Theorem, which cannot be obtained by the Cauchy integral formula. It is known that $\bar{\partial}$ question is a very important part in the modern theory of partial differential equation and a powerful tool in the modern mathematics. It might be more beneficial to give readers of this book some opportunity to feel the flavor of modern mathematics, which, therefore, has become the author's further consideration in designing the book. A series of theorems are prove by applying the solution of the $\bar{\partial}$ question in the third chapter of the book, with Mittag-Leffler Theorem (§3.4 Theorem 6, Chapter 3) in particular. By doing this, the proof of the theorems becomes very simple and, at the same time, the power of the $\bar{\partial}$ question is fully demonstrated. Traditional textbooks used to focus so much on the discussion of the analytic function that some textbooks were titled as analytic function theory. With mathematics advancing so fast, we have come to realize today more than ever before that it is far from enough to discuss only analytic function. And it seems already outdated to have a textbook named analytic function theory and the like. The emergence and application of the $\bar{\partial}$ question is one of the examples to support the viewpoint of the author described above.

It is perhaps the first time to fine Pompeiu Theorem and the solution of one-dimensional $\bar{\partial}$ question in a college teaching material concerning complex analysis in China. The content that Mittag-Leffler Theorem is proved by the solution of the $\overline{\partial}$ question has been designed and included for the first time in the teaching material at the college level by the author, which stands as an example of successfully dealing with the classical conclusions. The same is true of another example in the same item that interpolation theorem by applying the solution of the $\overline{\partial}$ question (§3.4, Theorem 7, Chapter 3). It is believed that more textbook compilers will follow this kind of practice because the proof is greatly simplified and clarified. In the present book, uniformly estimations of the derivatives of all orders of holomorphic functions on compact set (§2.3, Theorem 6(2), Chapter 2) is given as another application of the Pompeiu Theorem, which, definitely alien to the traditional teaching material, is so profound a theorem that can hardly proved without applying the Pompeiu Theorem. The reason why this the theorem should appear in the second chapter of the book is that the theorem is to have more applications and, at the same time, readers may feel the power of the theorem though it is only a small display of the master's talent.

3. One of the basic theorem in the complex analysis is Poincaré-Koebe uniformization theorem, which is a programmatic theorem. The theorem states: Any simply connected Riemann surface is holomorphically equivalent to the one of the following three domains: unit disk, complex plane $\mathbb{C}$, extended complex plane $\mathbb{C}^*$. It is one of the most important and beautiful theorems in complex analysis, which, together with Abel Theorem, and Riemann-Roch Theorem make the three most important theorems in the classic Riemann surface which are rarely seen in mathematics. The proof of the Poincaré-Koebe Theorem is well beyond the basic courses at the college level and cannot be rendered here but the theorem itself must be described and its significance emphasized here in the college textbook. It is, therefore, illustrated in the §3.3, Chapter 3. Besides a variety of applications, the most important significance of the theorem is that the theorem has established the important positions of unit disk, complex plane $\mathbb{C}$ and extended complex plane C^* in complex analysis. The practice of investigation and research

in these three domains has become the most important component in the study of complex analysis, which is consequently treated as something programmatic.

Group of holomorphic automorphisms of the three mentioned domains are given here in this book. the group of holomorphic automorphisms of the unit disk and group of holomorphic automorphisms of C and C* are given respectively in §2.5, Chapter 2 and §3.3, Chapter 3. Group of holomorphic automorphisms of an domain is extremely important because the properties of some analysis are decided to a great extent by the group of holomorphic automorphisms. In this book, an example is cited by using the group of holomorphic automorphisms of the unit disk to determine the Poisson kernel. Another purpose to emphasize the three group of holomorphic automorphisms is that they are simplest but most important Lie groups. These Lie groups and their applications are just introduced for readers' preview though it is unnecessary to discuss the strict concept of Lie groups in a textbook developed for undergraduates. With these three simple Lie groups moderately stressed for our readers today, they may have a feeling of meeting a fairly familiar friend tomorrow when they begin to study the strict concept of Lie groups. Having had something about Lie groups, they will be greatly benefited when it seems easier for them to understand and master the very abstract definition of Lie groups. It is out of the same consideration that the groups of holomorphic automorphisms of unit ball and bidisk in $\mathbb{C}^2$ are given and used to prove the classic Poincaré theorem in several complex variables. Then the reader have two simple and concrete examples of Lie groups more.

In Chapter 2 and 3, we gave the group of holomorphic automorphisms of three domains which indicate the relation between complex analysis and algebra. In Chapter 5, however, the complex geometry we discuss indicates the relation between complex analysis and geometry.

Poincaré-Koebe theorem can be rendered in a more accurate way:

(1) Any simply connected open hyperbolic Riemann surface is con-

formally equivalent to the unit disk;

(2) Any simply connected open parabolic Riemann surface is conformally equivalent to the complex place $\mathbb{C}$;

(3) And any simply connected closed Riemann surface is conformally equivalent to the extended complex plane $\mathbb{C}^*$.

In Chapter 5, we establish the geometry of these three domains, we equip the hyperbolic metric, i.e., the Poincaré metric on the unit disk; we equip the parabolic metric, i.e., the Euclidean metric on $\mathbb{C}$; we equip the elliptic metric, i.e., spherical metric on $\mathbb{C}^*$. Then a simple discussion of these three complex geometrices are followed. In my understanding, these material belong to the basic part of complex analysis.

4. We mentioned above the relation between complex analysis and geometry, algebra. This is a good example to demonstrate the unity of mathematics. The geometry mentioned above is differential geometry, the complex geometry, not the elementary geometry. Of course, we may use the elementary geometry to prove some elementary theorems in complex plane. But it is within the content for high school and, therefore, is reasonably excluded.

It is a good idea to explain the unity of mathematics by one example after another, especial in a textbook for a basic course. By designing in this way, we hope to help our readers develop this idea of the unity of mathematics, by these examples. And these are even more examples to follow.

In Chapter 2, §2.5, we discuss the Schwarz-Pick lemma. It endows the classical Schwarz Lemma with the differential geometrical meaning,. The lemma may state as follows: Any holomorphic mapping which maps unit disk into unit disk, makes the Poincaré distance of any two points in the unit disk nonincreesing. This is an elementary, natural and beautiful geometrical explanation of the classical Schwarz lemma. This is another good example to show the relation between complex analysis and differential geometry.

Using geometrical method to prove the famous Picard theorem in

Chapter 5 is another good example to explain the unity of mathematics. It is perhaps the first time that this kind of practice has appeared in a Chinese textbook of complex analysis. In 1938, Ahlfors established the important Ahlfors-Schwarz lemma, which was recorded in the history of mathematics because it marked the fact that the differential geometry had entered complex analysis. It is a lemma of historical importance. Starting from this lemma we prove the important Picard theorem in Chapter 5. The theorem, however, is not mentioned in traditional text-books because its proof is rather difficult, which asks for the use of the elliptic modular function. However, the proof can be simplified if we employ the method of differential geometry. And what is more? It is easy for the undergraduate student to understand. This, in turn, proves the importance of the Picard theorem. Therefore, we state and prove this theorem in Chapter 5. On one hand, we want the reader to know the important Picard theorem, on the other hand, we can expect the reader to face the power of the differential geometry when we use it. More important, let the reader know one more important example for the unity of mathematics. Between all the branches of mathematics, usually, it is the fact that there is something of each in the other.

The Poisson integral in Chapter 2 §2.6 is another example of the unity of mathematics. In the previous section, we already mentioned that Poisson integral formula comes from the group of holomorphic automorphisms of the unit disk. In this textbook we offer the Poisson integral formula from the group of holomorphic automorphisms, and avoid using the traditional analysis method, on purpose to let the reader feel the power of the group, especially that of the Lie group. By using a very simple idea of the group, we may obtain the important Poisson integral formula with ease. Moreover, in the point view of complex analysis, Poisson integral formula is a kind of integral representation; in the point view of partial differential equation, it is the solution of the Dirichlet problem of Laplace equation; in the point view of harmonic analysis, it is the Abel summability of the Fourier series of a function. The same

Poisson integral appears quite differently if viewed from different points of view. This observation is very important because it represents the permeability between different branches in mathematics. Usually from this point view, a result in one branch, will induce another result in another branch. The results of different branches are mutually inspired and mutually promoted. Of course, the Poisson integral is just one good simple example of this idea. In a textbook for an undergraduate course, we can only show some very simple examples to explain the idea. Actually, in mathematics, there are a lots of such examples. A typical one is the Riemann surface. In the point view of complex analysis, it is complex manifold of one dimension, more precisely, it is one-dimension Kähler manifold; but in the point view of algebraic geometry, it is an algebraic curve; further more in the point view of algebraic number theory, it is a field of algebraic function of one variable. The intersectional points of different branches of mathematics are usually the places of unusual vitality and importance.

5. In the last Chapter of this book, Chapter 6, we discuss and prove some results in several complex variables. It is absent in the traditional textbook of complex analysis. With the advances of the mathematics up to today, to take in a few results about several complex variables seems necessary and possible. The purpose to mention a few results about several cmplex variables is not to introduce some basic results of several complex variables to the reader. In doing so, it requires many pages. Of course it is unnecessary for basic textbook of complex analysis for undergraduate students. The purpose to state a few results about several complex variables is to let the reader understand one complex variable more deeply. Anyway, it is a textbook of one complex variable for undergraduate. For this aim, we select some theorems of several complex variables only which help the reader understand one complex variable much better. In this book, we select two theorems only. One is the Poincaré theorem, the other is the Hartogs theorem. We select these two theorems because their proofs are not very difficulty, and the

theorems themselves are truly fundamental. Moreover, there two theorems makes the reader to have a sound understanding of one complex variable. Poincaré theorem says: The unit ball and bidisk in $\mathbb{C}^2$ are not holomorphic equivalent. (It holds true for $\mathbb{C}^n$. For simplicity, we consider $\mathbb{C}^2$ only. The proof of the Theorem in $\mathbb{C}^n$ and the proof of the Theorem in $\mathbb{C}^2$ have no significant different). This Theorem tells us: The Riemann mapping theorem (topological equivalent implies holomorphic equivalent) holds true only in the case of one complex variable. Before that, in the calculus there are no such kind theorems. After that, in several complex variables, there are no such kind theorems again. Riemann mapping theorem holds only in one complex variable. Thus, it is a very profound theorem. As one of the three parts in complex analysis, the geometrical function theory can be established only in the case of one complex variable. Any attemption to extend the geometrical function theory from one complex variable to several complex variable will have to find the new ideas and new approaches. Another theorem for several complex variables which is introduced in this book is Hartogs theorem. This theorem says: Suppose $\Omega \subseteq \mathbb{C}^n$ is a domain, $n \geq 2, K$ is a compact subset is Ω, and $\Omega \backslash K$ is connected. If f is holomorphic on $\Omega \backslash K$, then f can holomorphic continued to Ω. This theorem tells us: As one of the three parts of complex analysis, Weierstrass theory on series, its essential part is the theory of Laurent series, but it usually does not exist in the case of several complex variables. Thus, how to describe the singularity by series, and how to define moromorphic function, etc. will ask for some new ideas and methods in the case of several complex variables. Hence, Laurent series happens only in the case of one complex variable. Before that, in calculus, it does not happen. After that, in several complex variables it does not happen either. Thus, it is a very profound theory. In conclusion, that is to say, as the starting points of the two out of the three parts of complex analysis, the Riemann mapping theorem and Laurent series is unique and unprecedented. Since this is a textbook for undergraduate students, it is more than enough to give and prove these two theorems for several complex variables. Otherwise, we will deviate from the topic of this book.

6. The fundamental content of the modern mathematics, which used to be absent in the traditional teaching material of complex analysis, are contained in the three appendices to the book.

The appendix to Chapter 2, partitions of unity, which is to be used in the proof of §2.3 Theorem 6 (2). It may be substituted by directly giving a function. However, the partitions of unity is so essential and so frequently used in the modern mathematics that it is worthwhile using some pages to write down the theorem and its proof for readers' future reference.

Chapter 4 has its appendix about the Riemann surface. The strict definition of the Riemann surface is offered here as a concerted effort with the content of §4.5 in the same chapter. Though it seems impossible to discuss too much about the theory of the Riemann surface in a college textbook, it is more than somewhat desirable to invite in the appendix and have it stay here. As an one-dimensional complex manifold, the theory of the Riemann surface is of such vital importance that the modern mathematics is hardly studied if this theory is not involved. To help our readers have some more preparations about the Riemann surface and complex manifold is also the author's original intention.

Curvature becomes the appendix to Chapter 5 for the purpose of the explanation of the why K defined by equation (1.2) in §5.1. Everybody knows that curvature is the core of differential geometry, playing a key role in the modern mathematics. Just for this reason, it proves to be more than helpful to repeat this significant concept though it has already studied in the differential geometry course at college.

The above three appendices are optional for classroom teaching activities.

It stands to reason that another purpose for the incorporation of the three appendices is to demonstrate the unity of mathematics.

Much has been illustrated on the difference between this book and traditional teaching material. In spite of all that has been said, most of the book is, however, traditional. The difference is, anyway, rather limited. But the author wants to express that it is just to this limited difference here in the book that the author has devoted his entire thought. The author sincerely wishes that all the readers, both teachers and students, would attach sufficient attention to the thought or the idea he has made extraodinarily painstaking efforts to highlight: that is of modernizing the teaching and learning of mathematics, and of emphasizing the unity of mathematics.

The intention of writing the book has been briefly introduced at some national academic conferences and evoked rather strong repercussions. However, to expect a book free from any error is just to expect the impossible. Undoubtedly, any criticism and suggestion from students, teachers and experts will be highly appreciated.

Sheng Gong

May, 2000, Beijing

Contents

CHAPTER I
CALCULUS

§ 1.1   A Glimpse of Calculus

The theory of functions of complex variable is to discuss calculus on the complex number field. Just like any extension of any branch in mathematics, some results can be obtained directly without any essential difficulty, and some results are happened in the complex number field only, it does not appear in the real number field. No doubt, the first part is important, but we usually concentrate our attention on the second part, because this part characterizes its essence of the matter.

In this chapter, we recall what is calculus briefly at first, then we observe what results in calculus can be extended to the complex number field. In the later chapters of this book we will discuss some important properties and results which are essentially different from the calculus in the real number field, and it happens on the complex number field only.

What is calculus? Calculus consists of three parts, namely, differential calculus, integral calculus and the fundamental theorem of calculus, the Newton-Leibniz formula, which indicates that differentiation and integration are inverse operations.

The following facts are well-known. If the function $y = f(x)$ is defined on (a, b), and $\lim\limits_{h \to 0} \frac{f(x+h)-f(x)}{h}$ exists at a point $x \in (a, b)$, then we call f is differentiable at the point x, and denote the limiting value by $\frac{\mathrm{d}f}{\mathrm{d}x}(x)$ or $f'(x)$, which is the derivative of f at point x; $\mathrm{d}f = f'(x)\mathrm{d}x$ is the differential of f at the point x. The function is differentiable on (a, b) if it is differentiable at any point $x \in (a, b)$. Suppose the function $y = f(x)$ is defined on $[a, b]$, we divide $[a, b]$ into n small intervals $[x_{i-1}, x_i]$ $(i = 1, \cdots, n)$, where $a = x_0 < x_1 < \cdots < x_n = b_i$. If ξ_i is any point in $[x_{i-1}, x_i]$ $(i = 1, 2, \cdots, n)$, and the length of every $[x_{i-1}, x_i]$ $(i = 1, 2, \cdots, n)$ tends to zero when $n \to \infty$, and $\lim\limits_{n \to \infty} \sum\limits_{i=1}^{n} f(\xi_i)(x_i - x_{i-1})$ exists, then we say that f is integrable on $[a, b]$, and denote it by $\int_a^b f(x)dx$. These are the basic definitions and starting points of calculus, and the derivative and the integral of a function have the clear geometric meaning. Derivative is the slope of the tangent of the curve $y = f(x)$ at the point (x, y). Integral is the area of the curved rectangular which is covered by $y = f(x)$ on $[a, b]$.

The ideas of differentiation and integration were established long time ago. The calculus became an independent branch of mathematics only when Newton and Leibniz proved the fundamental theorem of calculus. They pointed out that the differentiation and integration are inverse operations. This fundamental theorem of calculus has two equivalent forms.

Fundamental theorem of calculus (Differential form) Suppose function $f(x)$ is continuous on interval $[a, b]$, and x is a point in $[a.b]$. Let

$$\Phi(x) = \int_a^x f(t)\,dt, \qquad a \le x \le b$$

then $\Phi(x)$ is differentiable on $[a, b]$, and $\Phi'(x) = f(x)$, $d\Phi(x) = f(x)\,dx$. In other words, if the integration of $f(x)$ is $\Phi(x)$, then the differential of $\Phi(x)$ is $f(x)dx$.

Fundamental theorem of calculus (Integral form) If function $\Phi(x)$ is differentiable on interval $[a, b]$, and $\frac{d\Phi(x)}{dx}$ is a continuous function $f(x)$, then

$$\int_a^x f(t)\,dt = \Phi(x) - \Phi(a), \qquad a \le x \le b.$$

In other words, if the differential of $\Phi(x)$ is $f(x)dx$, then the integral of $f(x)$ is $\Phi(x)$.

From this theorem, we know that the integral is the inverse operation of differential. The properties of differential and integral are mutually correspondence, they become two sides of one event. For example, the formulas

$$\frac{d(f(x) \pm g(x))}{dx} = \frac{df(x)}{dx} \pm \frac{dg(x)}{dx}$$

and

$$\int (f(x) \pm g(x))\,dx = \int f(x)dx \pm \int g(x)dx$$

are correspondent to each other;

$$\frac{d}{dx}(fg) = f\frac{dg}{dx} + \frac{df}{dx}g$$

and the integration by parts

$$\int fg'\mathrm{d}x = fg - \int gf'\mathrm{d}x$$

are correspondent to each other. If $u = f(y)$, $y = g(x)$, then

$$\frac{\mathrm{d}f(g(x))}{\mathrm{d}x} = \frac{\mathrm{d}f}{\mathrm{d}y} \cdot \frac{\mathrm{d}y}{\mathrm{d}x}.$$

The above result and the formula

$$\int f(g(x))g'(x)\mathrm{d}x = \int f(y)\mathrm{d}y$$

are correspondent to each other, etc. Moreover, we have two mean-value theorems in calculus. One is the differential mean-value theorem: If $f(x)$ is differentiable on $[a, b]$, then there exists a point c in (a, b), such that

$$f(b) - f(a) = f'(c)(b - a).$$

The another one is the integral mean-value theorem: If $f(x)$ is continuous on $[a, b]$, then there exists a point ξ in $[a, b]$, such that

$$\int_a^b f(x)\mathrm{d}x = f(\xi)(b - a).$$

The differential mean-value theorem and the integral mean-value theorem are correspondent to each other. This is another example to show that the properties of differentiation and integration are correspondent to each other. The Taylor series is another example. We know already that the Taylor expansion of a function can be proved by the approach of differentiation and the remainder term of the series can be described by differentiation form; it can be proved by the approach of integration and the remainder term of the series can be desribed by integration form. Of course, we may give more examples, but it is unnecessary to go into more details.

In calculus, usually we discuss the elementary functions and their compositions. There are three kinds of elementary functions.

1. Power functions x^α, α is a real number; polynomials $a_0 + a_1 x + \cdots + a_n x^n$, where a_i $(i = 0, 1, \cdots, n)$ are constants; rational fraction $\frac{b_0 + b_1 x + \cdots + b_m x^m}{c_0 + c_1 x + \cdots + c_p x^p}$,

where b_i $(i = 0, 1, \cdots, m)$, c_i $(i = 0, 1, \cdots, p)$ are constants; and their inverse functions.

2. Trigonometric functions $\sin x$, $\cos x$, etc., and their inverse functions, for example, $\arcsin x$, $\arccos x$, etc.

3. Exponential functions $e^x, 2^x$, etc., and their inverse functions, $\ln x$, $\log_2 x$, etc.

We may understand that the Taylor expansion of $f(x)$ is the approximation of $f(x)$ by the first kind of elementary functions, the polynomials; and the Fourier expansion of $f(x)$ is the approximation of $f(x)$ by the second kind of elementary functions, $\sin nx$, $\cos nx$ $(n = 0, 1, 2, \cdots)$. But we have not the corresponding expansion of $f(x)$ by the third kind of elementary function, the exponential functions. The reason is that there exists an important formula, the Euler formula, we will mention it later. This formula tells us that the exponential function can be expressed in terms of trigonometric functions. The following Taylor series of some important elementary functions are well-known:

$$e^x = 1 + \frac{x}{1!} + \frac{x^2}{2!} + \frac{x^3}{3!} + \cdots, \tag{1.1}$$

$$\sin x = \frac{x}{1!} - \frac{x^3}{3!} + \frac{x^5}{5!} - \frac{x^7}{7!} + \cdots, \tag{1.2}$$

$$\cos x = 1 - \frac{x^2}{2!} + \frac{x^4}{4!} - \frac{x^6}{6!} + \cdots, \tag{1.3}$$

$$\ln(1 + x) = x - \frac{x^2}{2} + \frac{x^3}{3} - \cdots, \qquad -1 < x \leq 1, \tag{1.4}$$

$$(1 + x)^r = 1 + rx + \frac{r(r-1)}{2!}x^2 + \frac{r(r-1)(r-2)}{3!}x^3 + \cdots,$$
$$|x| < 1, \ r \text{ is a real number}, \tag{1.5}$$

etc.

We briefly recalled the calculus in above. To regard the calculus of one variable in this point of view, readers may refer to the textbook 《 Concise calculus 》 (S. Gong and S.L. Zhang [1]) for more detail.

For the calculus of the high dimensions, it has the corresponding three parts: differentiation, integration, and the fundamental theorem of calculus in high dimensions which points out the connection between differentiation and integration. In the part of differentiation, we have partial derivatives, total differential, and the Jacobi matrix which is corresponding the derivative in one variable. In the part of integration, we have multiple integration, line integral and surface integral, etc. All of them are natural extensions of the differentiation and integration from one dimension to high dimension. It is easy to list the corresponding theorems, but we omit it. For the third part, we have to say more. In high dimension, what is the fundamental theorem of calculus? How to describe the connection between differentiation and integration? The answer is: the Green formula, the Stokes formula and the Gauss formula altogether form the fundamental theorem of calculus in high dimensions and they describe the connection between differentiation and integration in high dimensions.

Green formula   If D is a closed domain which is bounded by a closed curve L in plane Oxy, functions $P(x, y)$ and $Q(x, y)$ have the continuous partial derivatives of order one in D, then

$$\oint_L P \, dx + Q \, dy = \iint_D \left(\frac{\partial Q}{\partial x} - \frac{\partial P}{\partial y} \right) dx \, dy. \tag{1.6}$$

Stokes formula   If surface Σ in 3-dimensional space is bounded by a closed curve L, functions $P(x, y, z)$, $Q(x, y, z)$ and $R(x, y, z)$ have the continuous partial derivatives of order one in Σ, then

$$\oint_L P \, dx + Q \, dy + R \, dz = \iint_\Sigma \left(\frac{\partial R}{\partial y} - \frac{\partial Q}{\partial z} \right) dy \, dz$$

$$+ \left(\frac{\partial P}{\partial z} - \frac{\partial R}{\partial x} \right) dz \, dx + \left(\frac{\partial Q}{\partial x} - \frac{\partial P}{\partial y} \right) dx \, dy. \tag{1.7}$$

Gauss formula   If V is a closed domain in 3-dimensional space which is bounded by a closed surface Σ, the functions $P(x, y, z), Q(x, y, z)$ and $R(x, y, z)$ have the continuous partial derivatives of order one in V, then

$$\oiint_\Sigma P \, dy \, dz + Q \, dz \, dx + R \, dx \, dy = \iiint_V \left(\frac{\partial P}{\partial x} + \frac{\partial Q}{\partial y} + \frac{\partial R}{\partial t} \right) dV. \tag{1.8}$$

These three formulas describe the relations between the integral on the boundary and the integral on the domain. If we use exterior differential form, then we may unify these three formulas into one formula, and it is called Stokes formula again. It needs a lots of pages to state the exterior differential form rigorously. The readers may find it in some textbooks of modern calculus. The textbook 《 Concise calculus 》 was written in this point of view. Here we briefly and formally introduce the exterior differential form in Euclidean space of three dimension. (cf. S. Gong and S.L. Zhang [1].)

Denote the exterior product of differentials dx and dy by $dx \wedge dy$, it obeys the following rules:

(1) $dx \wedge dx = 0$, the product of the differential and itself is zero.

(2) $dx \wedge dy = -dy \wedge dx$, the product of two different differentials changes sign if the order is changed.

Of course, we may regard (1) as a consequence of (2). Multiplying functions on exterior products, and adding altogether, it constructs an exterior differential form. For example, if P, Q, R, A, B, C and H are functions of x, y and z, then

$$P \, dx + Q \, dy + R \, dz$$

is exterior differential form of order one (It is the same as the differential form in the original sense because the exterior product does not appear in the form);

$$A \, dx \wedge dy + B \, dy \wedge dz + C \, dz \wedge dx$$

is the exterior differential form of order two;

$$H \, dx \wedge dy \wedge dz$$

is the exterior differential form of order three.

The functions P, Q, R, A, B, C and H are called the coefficients of the exterior differential forms.

We may define the exterior differential operator d for the exterior differential form ω. The definition is as follows.

For exterior differential form of order zero, the function f, we define

$$df = \frac{\partial f}{\partial x} \, dx + \frac{\partial f}{\partial y} \, dy + \frac{\partial f}{\partial z} \, dz,$$

it is the total differential operator in the original sense. For exterior differential form of order one $\omega = P dx + Q dy + R dz$, we define

$$d\omega = dP \wedge dx + dQ \wedge dy + dR \wedge dz.$$

It means, we differentiate P, Q and R first, then do exterior product with dx, dy and dz respectively. It is easy to verify by the rules of exterior product,

$$d\omega = \left(\frac{\partial R}{\partial y} - \frac{\partial Q}{\partial t}\right) dy \wedge dz + \left(\frac{\partial P}{\partial z} - \frac{\partial R}{\partial x}\right) dz \wedge dx + \left(\frac{\partial Q}{\partial x} - \frac{\partial P}{\partial y}\right) dx \wedge dy.$$

For exterior differential form of order two $\omega = Ady \wedge dz + Bdz \wedge dx + Cdx \wedge dy$, we define

$$d\omega = dA \wedge dy \wedge dz + dB \wedge dz \wedge dx + dC \wedge dx \wedge dy$$
$$= \left(\frac{\partial A}{\partial x} + \frac{\partial B}{\partial y} + \frac{\partial C}{\partial t}\right) dx \wedge dy \wedge dz.$$

For exterior differential form of order three $\omega = Hdx \wedge dy \wedge dz$, we define

$$d\omega = dH \wedge dx \wedge dy \wedge dz.$$

Obviously, it equals zero. If we make a rule that $ddx = ddy = ddz = 0$, then the exterior differential operator is the same as the differential operator in original sense, it operates to each term, then operates to each factor of the term and keep the other factors, then adds all these terms. The only difference is that the product is exterior differential product. We have the following important **Poincaré lemma**: If ω is an exterior differential form, its coefficients have continuous partial derivative of order two, then $dd\omega = 0$. The inverse Poincaré lemma holds. If ω is an exterior differential form of order p, and $d\omega = 0$, then there exists an exterior differential form α of order $p - 1$, such that $\omega = d\alpha$. By the above preparations, the Green formula, the Stokes formula and the Gauss formula can be unified as one formula

$$\int_{\partial \Sigma} \omega = \int_{\Sigma} d\omega, \tag{1.9}$$

where ω is the exterior differential form; and $d\omega$ is the exterior differential of ω; Σ is the domain of integration of $d\omega$, it is a closed domain; $\partial \Sigma$ is the boundary of Σ, $\int$ means the multiple integral, the multiplicity of the integral equals the dimension of the domain of integration. In fact, if ω is an exterior differential form of order zero, (1.9) reduces to the Newton-Leibniz formula; if ω is exterior differential form of order one, and the domain is planer, (1.9) reduces to the Green formula; if ω is exterior differential form of order one, and the domain is a surface in three dimensional space, (1.9) reduces to the

Stokes formula; if ω is exterior differential form of order two, (1.9) reduces to the Gauss formula. The formula (1.9) really described some cancellations between the differentiation and integration in three dimension. Actually (1.9) holds true for the high dimensions when the dimension is higher than three. Moreover, it holds true for differential manifold in general. Thus (1.9) is the fundamental theorem of calculus in high dimension. In certain sense, it is a top and the end of the calculus.

Of course, to recall calculus in above is rough. But we only need to declare the main idea, and need not to give more details.

The theory of functions of complex variable is the calculus on the complex number field. It is the success of the calculus. This is why we use formula (1.9) as the starting point of this book.

§ 1.2   Complex Number Field, Extended Complex Plane and Spherical Representation

All complex numbers form the field of complex numbers, it is an extension of the real number field.

In elementary algebra, we know that the imaginary number i has the property $i^2 = -1$. Combining i and two real number α, β by addition and multiplication, we obtain a complex number $\alpha + i\beta$, where α and β are the real part and imaginary part of the complex number respectively. Denote $\operatorname{Re} a = \alpha$, $\operatorname{Im} a = \beta$ if $a = \alpha + i\beta$. Two complex numbers are equal if and only if the real parts of the complex numbers are equal, and the imaginary parts of the complex numbers are equal. The arithmetic of complex numbers are

$$(\alpha + i\beta) \pm (\gamma + i\delta) = (\alpha \pm \gamma) + i(\beta \pm \delta),$$

$$(\alpha + i\beta)(\gamma + i\delta) = (\alpha\gamma - \beta\delta) + i(\alpha\delta + \beta\gamma),$$

$$\frac{\alpha + i\beta}{\gamma + i\delta} = \frac{(\alpha + i\beta)(\gamma - i\delta)}{(\gamma + i\delta)(\gamma - i\delta)} = \frac{(\alpha\gamma + \beta\delta) + i(\beta\gamma - \alpha\delta)}{\gamma^2 + \delta^2}$$

if $\gamma + i\delta \neq 0$.

The complex number $\alpha - i\beta$ is the conjugate complex number of the complex number $a = \alpha + i\beta$, and is denoted by $\bar{a}$. Hence

$$\operatorname{Re} a = \frac{a + \bar{a}}{2}, \qquad \operatorname{Im} a = \frac{a - \bar{a}}{2i},$$

$$\overline{a+b} = \bar{a} + \bar{b}, \quad \overline{ab} = \bar{a} \cdot \bar{b}, \quad \overline{\left(\frac{a}{b}\right)} = \frac{\bar{a}}{\bar{b}},$$

and $a\bar{a} = \alpha^2 + \beta^2$. We denote $a\bar{a}$ by $|a|^2$, and $|a| = \sqrt{\alpha^2 + \beta^2}$ is the absolute value of a. Obviously

$$|a| \geq 0, \quad |ab| = |a| \cdot |b|, \quad \left|\frac{a}{b}\right| = \frac{|a|}{|b|}, \quad b \neq 0,$$

$$|a \pm b|^2 = |a|^2 + |b|^2 \pm 2\text{Re}\,(a\bar{b}), \quad |a + b| \leq |a| + |b|,$$

etc.

If the rectangular coordinates system in the place is given, the complex number $a = \alpha + i\beta$ can be represented as a point with coordinates (α, β). The first coordinate axis is the real axis, and the second coordinate axis is the imaginary axis, and the plane is the complex plane, denoted by $\mathbb{C}$.

A complex number can be expressed as a point. It can be expressed as a vector starting from the origin and ending at the point also. We use same character a to express the complex number, the point and the vector.

If the vector is obtained by moving a vector parallel, it identify with the original vector as usually. The addition of two complex numbers is the addition of two vectors, and the formulas of complex numbers have their geometrical meaning. For example, if a and be are two vectors, then the formula $|a + b| \leq |a| + |b|$ means the sum of two sides of a triangle is greater or equal to the third side, etc.

We may use the polar coordinate to represent a complex number, $a = \alpha + i\beta = r(\cos \varphi + i \sin \varphi)$. Obviously, $r = |a|$, it is the **modulus** of complex number. φ is called the **argument** of complex number. If

$$a_1 = r_1(\cos \varphi_1 + i \sin \varphi_1), \qquad a_2 = r_2(\cos \varphi_2 + i \sin \varphi_2),$$

then

$$
\begin{aligned}
a_1 a_2 &= r(\cos \varphi + i \sin \varphi) \\
&= r_1 r_2 (\cos \varphi_1 + i \sin \varphi_1)(\cos \varphi_2 + i \sin \varphi_2) \\
&= r_1 r_2 \big(\cos(\varphi_1 + \varphi_2) + i \sin(\varphi_1 + \varphi_2) \big).
\end{aligned}
$$

We have $r = r_1 r_2$, $\varphi = \varphi_1 + \varphi_2$. The argument of complex number is not unique, since $\varphi + 2k\pi$, k is any integers, is an argument again. We denote the argument of a by $\operatorname{Arg} a$. In particular, if $0 \le \varphi < 2\pi$, we call it as principal argument, and denote it by $\arg a$.

If $z = x + iy$, a is a fix complex number, r is a non-negative real number, then $|z - a| = r$ describe a circle centered at a with radius r; $|z - a| < r$ means a disc centered at a with radius r, denote this disc by $D(a, r)$. Similarly, $\operatorname{Im} z > 0$ means the upper half plane, $\operatorname{Re} z > 0$ means the right half plane, etc.

How to treat point at infinity for the complex plane $\mathbb{C}$ when we introduce the coordinate system in the plane? In theory of functions of complex variable, we extend $\mathbb{C}$ by adding one point, the point at infinity, denote it by ∞. For any finite complex number $a \in \mathbb{C}$, $a + \infty = \infty + a = \infty$. For any $b \ne 0$, $b \cdot \infty = \infty \cdot b = \infty$, $\frac{a}{0} = \infty$ ($a \ne 0$), $\frac{b}{\infty} = 0$, etc. All the points in $\mathbb{C}$ and point at infinity ∞ form an extended complex plane, denote it by $\mathbb{C}^*$, $\mathbb{C}^* = \mathbb{C} \cup \{\infty\}$. In this book, $\mathbb{C}^*$ means the extended complex plane.

We establish a geometric model of the extended complex plane, in this model, every point in the extended complex plane has a concrete representation. This is spherical representation. We obtain it by stereographic projection.

Consider a unit sphere S^2 in 3-dimensional Euclidean space, its equation is $x_1^2 + x_2^2 + x_3^2 = 1$ (the rectangular coordinates of 3-dimensional space are x_1, x_2, x_3). For every point on S^2 except $(0,0,1)$, we may establish a one to one correspondence with a complex number

$$z = \frac{x_1 + i x_2}{1 - x_3}. \tag{2.1}$$

In fact, we have

$$|z|^2 = \frac{x_1^2 + x_2^2}{(1 - x_3)^2} = \frac{1 - x_3^2}{(1 - x_3)^2} = \frac{1 + x_3}{1 - x_3}$$

by (2.1). Hence

$$x_3 = \frac{|z|^2 - 1}{|z|^2 + 1}, \quad x_1 = \frac{z + \bar{z}}{1 + |z|^2}, \quad x_2 = \frac{z - \bar{z}}{1 + |z|^2}. \tag{2.2}$$

Let point at infinity correspond $(0,0,1)$, then we complete the one to one correspondence between the points of S^2 and the points of the extended complex plane $\mathbb{C}^*$. Thus we use the sphere S^2 as the reprentative of the extended complex plane $\mathbb{C}^*$. We call the sphere S^2 as **Riemann sphere**. Of course, the semi-sphere $x_3 < 0$ corresponds the unit disc $|z| < 1$, and the semi-sphere $x_3 > 0$ corresponds the exterior part of the unit disc $|z| > 1$, etc.

If x_1-axis is the real axis, x_2-axis is the imaginary axis of a complex plane, (2.1) has clear geometric meaning.

Let $z = x + i\,y$, then

$$x : y : -1 = x_1 : x_2 : x_3 - 1$$

by (2.1). This means that $(x, y, 0), (x_1, x_2, x_3), (0, 0, 1)$ are collinear. Thus this correspondence is a central projection with center $(0,0,1)$, it projects the points on S^2 onto the points on $\mathbb{C}^*$. This projection is **stereographic projection**. In spherical representation, nothing is special for the point at infinity.

§ 1.3   Complex Differentiation

Just like in calculus, we may define the complex valued function $w = f(z)$ on the field of complex numbers, where z, w are complex numbers. We assume that $f(z)$ is single valued. We may use $\varepsilon - \delta$ language to define the limit of a function,

$$\lim_{z \to a} f(z) = A,$$

as follows: for any $\varepsilon > 0$, there exists a positive number δ, such that $|f(z) - A| < \varepsilon$ holds for all z in $|z - a| < \delta$. $f(z)$ is continuous at $z = a$ if $\lim_{z \to a} f(z) = f(a)$.

Just like in calculus, we may define the open set, closed set, connected set and compactness, etc. A curve in complex plane is a continuous complex valued function $\gamma(t)$ defined on the interval $[\alpha, \beta]$, i.e., $\gamma(t) = x(t) + i\,y(t)$, $\alpha \le t \le \beta$, where $x(t)$ and $y(t)$ are continuous real valued functions. $\gamma(\alpha), \gamma(\beta)$ are end points of the curve $\gamma(t)$. The curve $\gamma(t)$ is a closed curve if $\gamma(\alpha) = \gamma(\beta)$. The direction of the curve is the direction as t increasing. The curve is said to be smooth if $\gamma'(t)$ exists and is continuous on $[\alpha, \beta]$. $\gamma(t)$ is a piecewise smooth curve if $\gamma'(t)$ is continuous on $[\alpha, \beta]$ except finite number of points, and at every exceptional point, the left and the right derivatives exist. The piecewise curve is rectifiable. $\gamma(t)$ is a simple curve or a Jordan curve if $\gamma(t_1) = \gamma(t_2)$ implies $t_1 = t_2$. $\gamma(t)$ is a simple closed curve or a Jordan closed curve if it is simple and closed.

A domain D in complex plane is a set which satisfies the following two conditions:

(1) D is an open set;

(2) D is connected, that is, we may join any two points in D by a curve in D.

The following fact is intuitive, but its proof of this fact is complicated. We state here without proof.

Jordan theorem Any simple closed curve γ divides the complex plane into two parts, one is bounded, the interior part bounded by γ; and the other one is unbounded, the exterior part bounded by γ, and γ is the common boundary of these two domains.

Denote the boundary of D by ∂D. D is **simply connected** if the interior part of any simple closed curve in D is contained in D. A domain is **multiple connected** if it is not simply connected. A two connected domain is bounded by two mutually non-intersecting closed Jordan curves. A n connected domain is bounded by n mutually non-intersecting closed Jordan curves. These closed curves may degenerate to a singe point or a single slit. Moreover, just like in the real variable case, we may prove Heine-Borel theorem, Bolzano-Weierstrass theorem, etc. Here we state these two theorem and omit the proofs.

Heine-Borel theorem If A is a compact set, G is an open covering of A, then we may choose a finite open covering of A from G.

Bolzano-Weierstrass theorem Any infinite set has at least one limiting point.

Now we consider the derivative of complex valued function $w = f(z)$ of complex variable z. It is natural to consider $\lim\limits_{h \to 0} \frac{f(z+h)-f(z)}{h}$, where h is a complex number. $f(z)$ is differentiable at point z if the limit exists, in other words, the limiting values are equal along any path when $h \to 0$. It is the derivative of $f(z)$ at z, and is denoted by $\frac{\mathrm{d}f}{\mathrm{d}z}$ or $f'(z)$. $f(z)$ is an **analytic function** or a **holomorphic function** on its domain of definition if it is differentiable at every point in this domain. This definition is the same as the definition of derivative in calculus. Hence the formulas about the arithmetic operators and the derivative of composite functions are the same as the corresponding formulas in calculus. It is easy to derive. However, the complex derivative is considered on the complex plane, there are something special.

If

$$f(z) = u(z) + \mathrm{i}\,v(z) = u(x,y) + \mathrm{i}\,v(x,y)$$

is differentiable at point $z_0 = x_0 + \mathrm{i}\,y_0$, then

$$\lim_{z \to z_0} \frac{f(z) - f(z_0)}{z - z_0} = f'(z_0)$$

exists and equals when $z \to z_0$ along any path. In particular, if $z \to z_0$ along a path parallel to x-axis, $z = x + iy_0$, $x \to x_0$, then

$$f'(z_0) = \lim_{x \to x_0} \left[\frac{u(x, y_0) - u(x_0, y_0)}{x - x_0} + i \frac{v(x, y_0) - v(x_0, y_0)}{x - x_0} \right]$$
$$= u_x(x_0, y_0) + i v_x(x_0, y_0),$$

if $z \to z_0$ along a path parallel to y-axis, $z = x_0 + iy$, $y \to y_0$, then

$$f'(z_0) = \lim_{y \to y_0} \left[\frac{u(x_0, y) - u(x_0, y_0)}{i(y - y_0)} + \frac{v(x_0, y) - v(x_0, y_0)}{y - y_0} \right]$$
$$= v_y(x_0, y_0) - i u_x(x_0, y_0),$$

where u_x, u_y, v_x, v_y denote the partial derivatives of u and v with respect to x, y respectively. Comparing the real part and imaginary part of two above equalities, we have

$$u_x = v_y, \qquad u_y = -v_x. \tag{3.1}$$

(3.1) can be rewritten as

$$\frac{\partial f}{\partial x} = -i \frac{\partial f}{\partial y}. \tag{3.2}$$

The equations (3.1) or (3.2) are called the **Cauchy-Riemann equations**, in short, C-R equations. C-R equations are necessary but not sufficient condition for differentiability. For example,

$$f(z) = f(x + iy) = \sqrt{|xy|}$$

satisfy C-R equations at $z = 0$, but it is non-differentiable at $z = 0$. Let $x = \alpha t$, $y = \beta t$,

$$\frac{f(z) - f(0)}{z - 0} = \frac{f(z)}{z} = \frac{\sqrt{|\alpha\beta|}}{\alpha + i\beta}$$

The limit values are different when $z \to 0$ along different paths. But we have the following theorem.

    Theorem 1   The function $f(z) = u + iv$ is holomorphic in a domain D if and only if u, v have continuous partial derivatives of order one and satisfy the C-R equations (3.1).

    Proof   Necessity. If $f(z)$ is differentiable at $z = z_0$, then it was proved that it satisfies the C-R equations (3.1). In §2.3 of Chapter II, we will prove that the derivative of a holomorphic function is holomorphic. Hence $f' = u_x + i v_x = v_y - i u_y$ is a continuous function.

Sufficiency. If u, v have continuous partial derivatives of order one at point $z_0 = x_0 + i y_0$, and satisfy the C-R equations. Let $\alpha = u_x(x_0, y_0)$, $\beta = v_x(x_0, y_0)$, then

$$u(x, y) - u(x_0, y_0) = \alpha(x - x_0) - \beta(y - y_0) + \varepsilon_1(|\Delta z|),$$

$$v(x, y) - v(x_0, y_0) = \beta(x - x_0) + \alpha(y - y_0) + \varepsilon_2(|\Delta z|),$$

where $|\Delta z| = \sqrt{(x - x_0)^2 + (y - y_0)^2}$, $\varepsilon_1, \varepsilon_2$ satisfy

$$\lim_{|\Delta z| \to 0} \frac{\varepsilon_1(|\Delta z|)}{|\Delta z|} = \lim_{|\Delta z| \to 0} \frac{\varepsilon_2(|\Delta z|)}{|\Delta z|} = 0.$$

Multiplying the second equality by i and adding to the first equality, we have

$$f(z) - f(z_0) = (\alpha + i\beta)(z - z_0) + \varepsilon_1(|\Delta z|) + i \varepsilon_2(|\Delta z|).$$

Dividing $z - z_0$ on both sides, we get

$$\frac{f(z) - f(z_0)}{z - z_0} - (\alpha + i\beta) = \frac{\varepsilon_1(|\Delta z|) + i \varepsilon_2(|\Delta z|)}{z - z_0}.$$

Hence

$$\lim_{z \to t_0} \frac{f(z) - f(z_0)}{z - z_0} = \alpha + i\beta,$$

it is

$$f'(z_0) = u_x(x_0, y_0) + i v_x(x_0, y_0).$$

We have proved the theorem.

In the §2.3 of Chapter II, we will proved that: if $f(z) = u + i v$ is holomorphic in D, then $f'(z)$ and $f''(z)$ are holomorphic in D, and hence the partial derivatives of order two are continuous, the mixed partial derivatives of order two $\frac{\partial^2 v}{\partial x \partial y}$ and $\frac{\partial^2 v}{\partial y \partial x}$ are equal. By C-R equations, we have

$$\frac{\partial^2 u}{\partial x^2} = \frac{\partial^2 v}{\partial x \partial y}, \qquad \frac{\partial^2 u}{\partial y^2} = -\frac{\partial^2 v}{\partial y \partial x},$$

hence

$$\frac{\partial^2 u}{\partial x^2} + \frac{\partial^2 u}{\partial y^2} = 0.$$

Similarly, we have

$$\frac{\partial^2 v}{\partial x^2} + \frac{\partial^2 v}{\partial y^2} = 0.$$

These equations are Laplace equation. It is one of the three most important partial differential equations, the typical equation of elliptic partial differential equation, and denoted it by

$$\Delta u = \frac{\partial^2 u}{\partial x^2} + \frac{\partial^2 u}{\partial y^2} = 0,$$

where

$$\Delta = \frac{\partial^2}{\partial x^2} + \frac{\partial^2}{\partial y^2}.$$

A function u is harmonic function if u satisfies $\Delta u = 0$. The real part and imaginary part of a holomorphic function $f = u + iv$ are harmonic functions.

Since $z = x + iy$, $\bar{z} = x - iy$, we have $x = \frac{1}{2}(z + \bar{z})$ and $y = \frac{-1}{2}i(z - \bar{z})$. A function $f(x, y)$ can be considered as a function of z and $\bar{z}$, and regard z and $\bar{z}$ as independent variables (Actually they are mutually conjugate, but we do not concern it). If we use the differential rule, we have

$$\frac{\partial f}{\partial z} = \frac{1}{2}\left(\frac{\partial f}{\partial x} - i\frac{\partial f}{\partial y}\right), \qquad \frac{\partial f}{\partial \bar{z}} = \frac{1}{2}\left(\frac{\partial f}{\partial x} + i\frac{\partial f}{\partial y}\right).$$

Using this notation, a function is holomorphic if and only if $\frac{\partial f}{\partial \bar{z}} = 0$. Roughly speaking, a holomorphic function is a function which is independent of $\bar{z}$ and dependent on z only. Thus a holomorphic function is a function of z exactly, so we do not call it as a complex valued function of two real variables. $\frac{\partial f}{\partial \bar{z}} = 0$ is equivalent to $\frac{\partial f}{\partial z} = \frac{\partial f}{\partial x}$. Using these notation, $\Delta = 4\frac{\partial}{\partial z}\frac{\partial}{\partial \bar{z}} = 4\frac{\partial}{\partial \bar{z}}\frac{\partial}{\partial z}$.

The complex derivative has an important property, the conformal property.

If a function $f(z)$ is holomorphic on a domain D, $z_0 \in D$, $f'(z_0) \neq 0$, $\gamma(t)$ $(0 \leq t \leq 1)$ is a smooth curve in D through z_0, and $\gamma(0) = z_0$, the angle between the tangent line of the curve at z_0 and the real axis is $\arg \gamma'(0)$, $f(z)$ maps $\gamma(t)$ onto a smooth curve $\sigma(t) = f(\gamma(t))$ through the point $w_0 = f(z_0)$, then $\sigma'(t) = f'(\gamma(t))\gamma'(t)$, $\sigma'(0) = f'(z_0) \cdot \gamma'(0)$, the angle between the tangent line of $\sigma(t)$ at point w_0 and the real axis is

$$\arg \sigma'(0) = \arg f'(z_0) + \arg \gamma'(0),$$

we have

$$\arg \sigma'(0) - \arg \gamma'(0) = \arg f'(z_0).$$

That means, the difference between the argument of the tangent line of $\sigma(t)$ at point w_0 and the argument of the tangent line of $\gamma(t)$ at point z_0 is $\arg f'(z_0)$, which is independent of $\gamma(t)$. Hence, if we have any two smooth curves $\gamma_1(t)$,

$\gamma_2(t)$ $(0 \leq t \leq 1)$, $\gamma_1(0) = \gamma_2(0) = z_0$, through the point z_0, holomorphic function $w = f(z)$ maps z_0 to $w_0 = f(z_0)$ and $\gamma_1(t)$, $\gamma_2(t)$ to $\sigma_1(t)$, $\sigma_2(t)$ respectively, then

$$\arg \sigma_2'(0) - \arg \gamma_2'(0) = \arg \sigma_1'(0) - \arg \gamma_1'(0).$$

It is

$$\arg \sigma_2'(0) - \arg \sigma_1'(0) = \arg \gamma_2'(0) - \arg \gamma_1'(0).$$

Thus the angle between $\gamma_1(t)$ and $\gamma_2(t)$ at point z_0 equals the angle between $\sigma_1(t)$ and $\sigma_2(t)$ at point $w_0 = f(z_0)$. Hence, the angle between two smooth curves through a point z_0 as well as its orientation is preserved under a holomorphic mapping $w = f(z)$ if $f'(z_0) \neq 0$. This is one conformal property of $f(z)$ at point z_0.

On the other hand, since

$$f'(z_0) = \lim_{z \to z_0} \frac{f(z) - f(z_0)}{z - z_0},$$

$f(z)$ maps a curve $\gamma(t)$ through z_0 to a curve $\sigma(t)$, then

$$\lim_{\substack{z \to z_0 \\ z \in \gamma}} \frac{|f(z) - f(z_0)|}{|z - z_0|} = \lim_{\substack{z \to z_0 \\ z \in \gamma}} \frac{|w - w_0|}{|z - z_0|} = |f'(z_0)|.$$

Thus the limit of the ratio of the distance between two image points and the distance of those two points is $|f'(z_0)|$, the magnitude of $f(z)$ at point z_0 which is independent of the curve γ. f maps any small triangle with one vertex z_0, to a small curved triangle, these two small triangles are approximately similar, i.e., these two differentiate triangles are similar. Combining these two properties in above, these are the conformal properties of the holomorphic function. Thus we call a holomorphic function in a domain D as **conformal mapping** (if $f'(z) \neq 0$). In Chapter IV, we will discuss it again.

§ 1.4 Complex Integration

If $f(t) = u(t) + \mathrm{i}\,v(t)$ is a complex valued function on a real interval $[a, b]$, then

$$\int_a^b f(t)\,\mathrm{d}t = \int_a^b u(t)\,\mathrm{d}t + \mathrm{i} \int_a^b v(t)\,\mathrm{d}t.$$

If γ is a piecewise differentiable arc, its equation is $z = z(t)$ $(a \leq t \leq b)$, $f(z)$ is defined and continuous on γ, then $f(z(t))$ is a continuous function on γ, we define

$$\int_{\gamma} f(z) \, dz = \int_a^b f(z(t)) z'(t) \, dt$$

as the integral of $f(z)$ along γ, which is invariant under the parameter transformation. If the increasing function $t = t(\tau)$ maps $\alpha \leq \tau \leq \beta$ to $a \leq t \leq b$, $t(\tau)$ is piecewise differentiable, then

$$\int_a^b f(z(t)) z'(t) \, dt = \int_a^b f(z(t(\tau))) z'(t(\tau)) \, dt = \int_{\alpha}^{\beta} f(z(t(\tau))) \frac{dz(t(\tau))}{d\tau} \, d\tau.$$

If we use the Riemann sum to define the line integral, we may get the same result. Hence we have the same properties as the line integral in calculus. For example,

$$\int_{-\gamma} f(z) \, dz = - \int_{\gamma} f(z) \, dz;$$

$$\int_{\gamma_1 + \gamma_2 + \cdots + \gamma_n} f(z) \, dz = \int_{\gamma_1} f(z) \, dz + \int_{\gamma_2} f(z) \, dz + \cdots + \int_{\gamma_n} f(z) \, dz$$

where, $\gamma, \gamma_1, \gamma_2, \cdots, \gamma_n$ are curves. Thus we need not to say more about the complex integration.

We have roughly discussed the complex differentiation and complex integration already. In calculus on the field of complex number, what is the corresponding results of the third part of calculus? In the complex plane, it is the Green formula in complex form. Here we use complex exterior differential form, and regard $z, \bar{z}$ as independent variables. We define the exterior product of differential form by

$$dz \wedge dz = 0, \quad d\bar{z} \wedge d\bar{z} = 0, \quad dz \wedge d\bar{z} = -d\bar{z} \wedge dz,$$

where

$$dz = dx + i \, dy, \quad d\bar{z} = dx - i \, dy.$$

Hence

$$\begin{aligned}
d\bar{z} \wedge dz &= (dx - i \, dy) \wedge (dx + i \, dy) \\
&= -i \, dy \wedge dx + i \, dx dy \\
&= 2 i \, dx \wedge dy = 2 i \, dA
\end{aligned}$$

where $\mathrm{d}A$ is the 2-dimensional area element. Just like the real variable case, we define the exterior differential form of order zero as a function $f(z, \bar{z})$; the exterior differential form of order one is $\omega_1 \mathrm{d}z + \omega_2 \mathrm{d}\bar{z}$, where ω_1, ω_2 are functions of $z, \bar{z}$; the exterior differential form of order two is $\omega_0 \mathrm{d}z \wedge \mathrm{d}\bar{z}$, where ω_0 is the function of $z, \bar{z}$. Define the exterior differential operator d on an exterior differential form as

$$\mathrm{d}\omega = \partial\omega \wedge \mathrm{d}z + \bar{\partial}\omega \wedge \mathrm{d}\bar{z},$$

where $\bar{\partial} = \frac{\partial}{\partial \bar{z}}$, $\partial = \frac{\partial}{\partial z}$. It is easy to prove that $\mathrm{dd}\omega = 0$ holds true for any exterior differential form ω. The Green formula in complex form is as follows.

Theorem 2 If $\omega = \omega_1 \mathrm{d}z + \omega_2 \mathrm{d}\bar{z}$ is an exterior differential form of order one in a domain Ω, where $\omega_1 = \omega_1(z, \bar{z})$, $\omega_2 = \omega_2(z, \bar{z})$ are differentiable functions of $z, \bar{z}$, if d is the exterior differential operator, $\mathrm{d} = \partial + \bar{\partial}$, $\partial = \frac{\partial}{\partial z}$, $\bar{\partial} = \frac{\partial}{\partial \bar{z}}$, $\partial\Omega$ is the boundary of Ω, then

$$\int_{\partial\Omega} \omega = \iint_{\Omega} \mathrm{d}\omega. \tag{4.1}$$

Proof If $\omega_1 = \xi_1 + \mathrm{i}\eta_1$, $\omega_2 = \xi_2 + \mathrm{i}\eta_2$ where $\xi_1, \eta_1, \xi_2, \eta_2$ are real valued functions, then

$$\begin{aligned}
\omega &= \omega_1 \mathrm{d}z + \omega_2 \mathrm{d}\bar{z} \\
&= (\xi_1 + \mathrm{i}\eta_1)(\mathrm{d}x + \mathrm{i}\,\mathrm{d}y) + (\xi_2 + \mathrm{i}\eta_2)(\mathrm{d}x - \mathrm{i}\,\mathrm{d}y) \\
&= \big((\xi_1 + \xi_2)\mathrm{d}x + (-\eta_1 + \eta_2)\mathrm{d}y\big) + \mathrm{i}\big((\eta_1 + \eta_2)\mathrm{d}x + (\xi_1 - \xi_2)\mathrm{d}y\big),
\end{aligned}$$

and

$$\begin{aligned}
\mathrm{d}\omega &= \partial(\omega_1 \mathrm{d}z + \omega_2 \mathrm{d}\bar{z}) + \bar{\partial}(\omega_1 \mathrm{d}z + \omega_2 \mathrm{d}\bar{z}) \\
&= \frac{\partial\omega_1}{\partial z}\mathrm{d}z \wedge \mathrm{d}z + \frac{\partial\omega_2}{\partial z}\mathrm{d}z \wedge \mathrm{d}\bar{z} + \frac{\partial\omega_1}{\partial \bar{z}}\mathrm{d}\bar{z} \wedge \mathrm{d}z + \frac{\partial\omega_2}{\partial \bar{z}}\mathrm{d}\bar{z} \wedge \mathrm{d}\bar{z} \\
&= \left(\frac{\partial\omega_1}{\partial \bar{z}} - \frac{\partial\omega_2}{\partial z}\right)\mathrm{d}\bar{z} \wedge \mathrm{d}z \\
&= \left[\frac{1}{2}\left(\frac{\partial}{\partial x} + \mathrm{i}\frac{\partial}{\partial y}\right)(\xi_1 + \mathrm{i}\eta_1) - \frac{1}{2}\left(\frac{\partial}{\partial x} - \mathrm{i}\frac{\partial}{\partial y}\right)(\xi_2 + \mathrm{i}\eta_2)\right]2\,\mathrm{i}\,\mathrm{d}A \\
&= \left[-\left(\frac{\partial\xi_1}{\partial y} + \frac{\partial\eta_1}{\partial x} + \frac{\partial\xi_2}{\partial y} - \frac{\partial\eta_2}{\partial x}\right) + \mathrm{i}\left(\frac{\partial\xi_1}{\partial x} - \frac{\partial\eta_1}{\partial y} - \frac{\partial\xi_2}{\partial x} - \frac{\partial\eta_2}{\partial y}\right)\right]\mathrm{d}A.
\end{aligned}$$

By Green formula (1.6), we have

$$\int_{\partial\Omega} (\xi_1 + \xi_2)\mathrm{d}x + (-\eta_1 + \eta_2)\mathrm{d}y = \iint_{\Omega} \left(-\frac{\partial}{\partial x}(\eta_1 - \eta_2) - \frac{\partial}{\partial y}(\xi_1 + \xi_2) \right)\mathrm{d}A$$

and

$$\int_{\partial\Omega} (\eta_1 + \eta_2)\mathrm{d}x + (\xi_1 - \xi_2)\mathrm{d}y = \iint_{\Omega} \left(\frac{\partial}{\partial x}(\xi_1 - \xi_2) - \frac{\partial}{\partial y}(\eta_1 + \eta_2) \right)\mathrm{d}A.$$

Hence (4.1) holds.

Actually, (4.1) holds true for the complex Euclidean space of high dimension, (4.1) holds true even for the complex manifold. This formula in general form is called Stokes formula again. The formula (4.1) is the start point of the next chapter.

§ 1.5   Elementary Functions

In calculus, the elementary functions consist three kinds of functions and its compositions. These three kinds of functions are:

(1) Power functions, polynomials, rational fractional functions and their inverse functions;

(2) Trigonometric functions and their inverse functions:

(3) Exponential functions and their inverse functions, the logarithm functions.

How to define these three kinds of functions in the complex number field? Some is easy, for example, the polynomials, we only need to change the real variable to complex variable. But some is not so easy, for example, $\sin z, e^z$, what is the meaning of $\sin z, e^z$ when z is a complex number? We need to define it in the complex number field. It should have the concrete and exact meaning and should coincide with the definition of the function in the real number field. A natural idea is to establish these definitions by series.

If y is a real number, then

$$e^y = 1 + \frac{y}{1!} + \frac{y^2}{2!} + \cdots + \frac{y^n}{n!} + \cdots .$$

It is natural to define

$$e^{iy} = 1 + \frac{(iy)}{1!} + \frac{(iy)^2}{2!} + \cdots + \frac{(iy)^n}{n!} + \cdots$$

$$= 1 + \frac{iy}{1!} - \frac{y^2}{2!} - \frac{iy^3}{3!} + \frac{y^4}{4!} + \cdots$$

$$= \left(1 - \frac{y^2}{2!} + \frac{y^4}{4!} - \cdots\right) + i\left(\frac{y}{1!} - \frac{y^3}{3!} + \frac{y^5}{5!} - \cdots\right).$$

But we know that

$$1 - \frac{y^2}{2!} + \frac{y^4}{4!} - \cdots = \cos y, \qquad \frac{y}{1!} - \frac{y^3}{3!} + \frac{y^5}{5!} - \cdots = \sin y.$$

Hence we have

$$e^{iy} = \cos y + i \sin y. \tag{5.1}$$

This is the well-known Euler formula.

It suggests us to define

$$e^z = e^x(\cos y + i \sin y) \tag{5.2}$$

for any complex number $z = x + iy$.

Then, (5.1) is the consequence of (5.2). (5.1) is a very important formula, which tells us that the exponential functions and the trigonometric functions can be represented by each other. From (5.1), we have

$$\cos y = \frac{e^{iy} + e^{-iy}}{2}, \qquad \sin y = \frac{e^{iy} - e^{-iy}}{2i}.$$

It suggests us to define

$$\cos z = \frac{e^{iz} + e^{-iz}}{2}, \qquad \sin z = \frac{e^{iz} - e^{-iz}}{2i} \tag{5.3}$$

for any complex number $z = x + iy$.

Of course, we may define $\tan z = \frac{\sin z}{\cos z}$, etc. From (5.3), we have $\cos iy = \operatorname{ch} y$, $\sin iy = i \operatorname{sh} y$ when y is real.

From definition (5.2), we may define the inverse function $\log z$ of e^z. It can be defined as follows. All the complex numbers w which satisfy $e^w = z$, w is the logarithm of z, and is denoted by $\operatorname{Log} z$. Similarly, we may define the inverse function $\arcsin z$, $\arccos z$ of $\sin z$, $\cos z$, etc.

For the power function z^α, the meaning is clear when α is an integer. For any complex number α, it is natural to define

$$w = z^\alpha = e^{\alpha \log z}. \tag{5.4}$$

The properties of the functions defined by (5.2), (5.3) and (5.4) will discussed more in the later. However, we find that the three kinds of elementary functions are mutually not related in calculus, but in the complex number field they become one kind of elementary functions, the exponential functions and their inverse functions . The trigonometric functions and their inverse function can be represented by exponential functions and their inverse functions. The power functions and their inverse function can be represented by exponential functions and their inverse functions also. The key step to unify these three kinds of elementary functions in complex number field as one kind is the formula (5.1), the Eular formula. It is a very deep formula. For example, if $y = \pi$ in (5.1), then $e^{i\pi} = -1$. This formula related the four most important constants in mathematics, e, π, i and -1 in one identity. Moreover, the useful De Moivre formula

$$(\cos y + i \sin y)^n = \cos ny + i \sin ny$$

is a easy consequence of (5.1), etc.

Now we discuss the properties of these elementary functions.

We discuss the exponential function at first. By definition (5.2), we know that:

(1) The exponential function is non-zero, $e^z \neq 0$ since $|e^z| = e^x > 0$.

(2) For any $z_1 = x_1 + i y_1$, $z_2 = x_2 + i y_2$, we have $e^{z_1} e^{z_2} = e^{z_1 + z_2}$ since

$$e^{z_1} e^{z_2} = e^{x_1}(\cos y_1 + i \sin y_1)e^{x_2}(\cos y_2 + i \sin y_2)$$
$$= e^{x_1 + x_2}\left(\cos(y_1 + y_2) + i \sin(y_1 + y_2)\right) = e^{z_1 + z_2}.$$

(3) e^z has period $2\pi i$ since $e^{2\pi i} = 1$.

By (5.2), we have

$$u(x, y) = e^x \cos y, \qquad v(x, y) = e^x \sin y$$

and

$$u_x = v_y = e^x \cos y, \qquad u_y = -v_x = -e^x \sin y$$

are continuous functions on $\mathbb{C}$ where $e^z = u(x, y) + \mathrm{i}\, v(x, y)$. Hence e^z is holomorphic on $\mathbb{C}$ by Theorem 1, and

$$(e^z)' = u_x + \mathrm{i}\, v_x = e^x \cos y + \mathrm{i}\, e^x \sin y = e^z.$$

Thus we have

(4) e^z is holomorphic on $\mathbb{C}$, and $(e^z)' = e^z$.

The above discussions mentioned that the main properties of e^x in the real number field are preserved for the exponential functions in the complex number fields.

We regard $w = f(z)$ as a mapping from a domain in z-plane into a domain in w-plane. The mapping is univalent if the mapping is one to one. Now, we consider

(5) The univalent domain of e^z, i.e., find the domain where e^z is univalent. Suppose $z_1 = x_1 + \mathrm{i}\, y_1$, $z_2 = x_2 + \mathrm{i}\, y_2$ and $e^{z_1} = e^{z_2}$, thus

$$e^{x_1}(\cos y_1 + \mathrm{i} \sin y_1) = e^{x_2}(\cos y_2 + \mathrm{i} \sin y_2)$$

holds. It is $e^{x_1} e^{\mathrm{i}\, y_1} = e^{x_2} e^{\mathrm{i}\, y_2}$. Hence $x_1 = x_2$, $y_1 = y_2 + 2k\pi$, it implies $z_1 - z_2 = 2k\pi\mathrm{i}$, where k is an integer. The strips $2k\pi < y < 2(k+1)\pi$ $(k = 0, \pm1, \pm2, \cdots)$ are univalent domains of e^z. For example, $z = x + \mathrm{i}y$ $(0 < y < 2\pi)$, then e^z is an univalent mapping of this strip to $\mathbb{C}$ with a slit $\{z \mid z \geq 0\}$.

Next we discuss the inverse functions of the exponential functions, the logarithm functions.

A complex number w satisfying the equation $e^w = z$ is the logarithm of z, and denoted by $\mathrm{Log}\, z$. By the periodicity of exponential function, $\mathrm{Log}\, z$ is an infinity many valued function.

Let

$$z = re^{\mathrm{i}\theta}, \qquad w = u + \mathrm{i}v,$$

then

$$e^{u+\mathrm{i}v} = re^{\mathrm{i}\theta},$$

and hence $e^u = r$, $v = \theta + 2k\pi$, where k is any integer. We have

$$w = \log r + (\theta + 2k\pi)\,\mathrm{i},$$

or

$$w = \log |z| + \mathrm{i}\, \mathrm{Arg}\, z$$

where $\mathrm{Arg}\, z = \theta + 2k\pi$ is the argument of z.

The logarithm function has the property

$$\mathrm{Log}\,(z_1 z_2) = \mathrm{Log}\,z_1 + \mathrm{Log}\,z_2$$

since $\mathrm{Arg}\,(z_1 z_2) = \mathrm{Arg}\,z_1 + \mathrm{Arg}\,z_2$.

By the discussion on the exponential functions, we know that if D is $\mathbb{C}$ with a slit $\{z \mid z \geq 0\}$, and the argument of z is the principal argument $0 < \arg z < 2\pi$, then the function

$$w_k(z) = \log|z| + \mathrm{i}\,(\arg z + 2k\pi), \qquad k = 0, \pm1, \pm2, \cdots$$

is an univalent mapping of D onto the strip domain $E_k : 2k\pi < v < 2(k+1)\pi$, which is parallel to the real axis. All these $w_k(z)$ are the inverse functions of exponential function $z = e^w$, holomorphic on D and $w'_k = \frac{1}{z}$.

$$w_0(z) = \log|z| + \mathrm{i}\arg z$$

is the principal branch of $\mathrm{Log}\,z$, and we denote it by $\log z$,

$$\log z = \log|z| + \mathrm{i}\arg z.$$

For convenience, we let $-\pi < \arg z < \pi$ in some situations.

Next, we discuss the trigonometric functions. By the definition (5.3), we know that:

(1) $\cos z$, $\sin z$ are holomorphic on $\mathbb{C}$, and

$$(\sin z)' = \cos z, \qquad (\cos z)' = -\sin z.$$

(2) $\cos z$, $\sin z$ have the period 2π,

$$\sin(z + 2\pi) = \sin z, \qquad \cos(z + 2\pi) = \cos z.$$

(3) $\cos z$ is an even function, $\sin z$ is an odd function,

$$\sin(-z) = -\sin z, \qquad \cos(-z) = \cos z.$$

(4) Formulas of sum of angles

$$\sin(z_1 + z_2) = \sin z_1 \cos z_2 + \cos z_1 \sin z_2,$$
$$\cos(z_1 + z_2) = \cos z_1 \cos z_2 - \sin z_1 \sin z_2$$

are hold true.

(5) The fundamental relations of $\cos z$ and $\sin z$,

$$\sin^2 z + \cos^2 z = 1, \qquad \sin\left(\frac{\pi}{2} - z\right) = \cos z$$

are hold true.

(6) $\sin z = 0$ provided $z = k\pi$, $\cos z = 0$ provided $z = \frac{\pi}{2} + k\pi$, where $k = 0, \pm 1, \pm 2, \cdots$.

The above discussions mentioned that the main properties of $\sin x, \cos x$ in the real number field are preserved for the $\sin z, \cos z$ in the complex number field. But there are some differences between $\cos x, \sin x$ (x is real) and $\cos z, \sin z$ (z is complex).

(7) $|\sin z|$ and $|\cos z|$ are unbounded. By (4), we have

$$|\sin z|^2 = |\sin (x + \mathrm{i}\, y)|^2 = |\sin x \cos \mathrm{i}\, y + \cos x \sin \mathrm{i}\, y|^2$$
$$= |\sin x \operatorname{ch} y + \mathrm{i}\, \cos x \operatorname{sh} y|^2 = \operatorname{sh}^2 y + \sin^2 x.$$

It is an unbounded function. Similarly, $|\cos z|^2 = \operatorname{ch}^2 y - \sin^2 x$, it is an unbounded function too.

(8) The univalent domains of $\sin z$ and $\cos z$.

We consider $w = \cos z = \frac{e^{\mathrm{i}\, z} + e^{-\mathrm{i}\, z}}{2}$ at first. It is composed by three functions, $\xi = \mathrm{i}\, z$, $\zeta = e^{\xi}$, $w = \frac{1}{2}(\zeta + \frac{1}{\zeta})$. The first function is rotation, its mapping is univalent everywhere. The second function is univalent on a domain if and only if the domain in ξ-plane does not containing any two points ξ_1 and ξ_2 with $\xi_2 - \xi_1 = 2k\pi\mathrm{i}$, where k is any integer. The third function is univalent on a domain if and only if the domain in ζ plane does not containing any two points ζ_1 and ζ_2 with $\zeta_1\zeta_2 = 1$. In z-plane, the domain does not containing any two points z_1 and z_2 with $e^{\mathrm{i}\, z_1} \cdot e^{\mathrm{i}\, z_2} = 1$, that is $z_1 + z_2 = 2k\pi$, where k is any integer. Thus $\cos z$ is univalent on the strip $0 < \operatorname{Re} z < \pi$.

$\xi = \mathrm{i}\, z$ is an univalent mapping of $0 < \operatorname{Re} z < \pi$ onto $0 < \operatorname{Im} z < \pi$, $\zeta = e^{\xi}$ is an univalent mapping of $0 < \operatorname{Im} z < \pi$ onto the upper half plane $\operatorname{Im} \zeta > 0$. Finally $w = \frac{1}{2}(\zeta + \frac{1}{\zeta})$ is an univalent mapping of $\operatorname{Im} \zeta > 0$ onto w-plane with two slits on real axis: $-\infty < u \leq -1$ and $1 \leq u < +\infty$. Thus $w = \cos z$ is an univalent mapping of $0 < \operatorname{Re} z < \pi$ onto w-plane with two slits: $-\infty < u \leq -1$, $v = 0$ and $1 \leq u < +\infty$, $v = 0$.

By the same method, we may obtain the univalent domains of $\sin z, \tan z$, etc.

Next we discuss the inverse functions of trigonometric functions. We consider $w = \arccos z$, that is $\cos w = z$ at first.

We know that

$$\cos w = \frac{1}{2}(e^{\mathrm{i}w} + e^{-\mathrm{i}w}) = z.$$

This is a quadratic equation of $e^{\mathrm{i}w}$, it has two roots $e^{\mathrm{i}w} = z \pm \sqrt{z^2 - 1}$,

$$w = \arccos z = -\mathrm{i}\,\mathrm{Log}\,(z \pm \sqrt{x^2 - 1}) = \pm\mathrm{i}\,\mathrm{Log}\,(z + \sqrt{z^2 - 1}).$$

Hence $\arccos z$ is infinity many valued. It reflects the periodicity of $\cos w$. We may define $\arcsin z$ by $\frac{\pi}{2} - \arccos z$.

Finally, we discuss the power function.

By the definition (5.4),

$$z^\alpha = e^{\alpha\,\mathrm{Log}\,z} = e^{(a+\mathrm{i}b)(\log|z|+\mathrm{i}(\arg z+2k\pi))}$$
$$= e^{a\log|z|-b(\arg z+2k\pi)} \cdot e^{\mathrm{i}[b\log|z|+a(\arg z+2k\pi)]}$$

where $\alpha = a + \mathrm{i}b$, k is any integer. Let

$$\rho_k = e^{a\log|z|-b(\arg z+2k\pi)}, \qquad \theta_k = b\log|z| + a(\arg z + 2k\pi),$$

then

$$w = z^\alpha = \rho_k e^{\mathrm{i}\theta_k}, \qquad |w| = \rho_k.$$

Hence $w = z^\alpha$ is an infinity many valued function if $b \neq 0$. If $b = 0$, then α is a real number, and

$$w = z^\alpha = e^{a\log|z|}e^{\mathrm{i}(\arg z+2k\pi)a} = |z|^a e^{a(\arg z+2k\pi)\,\mathrm{i}},$$

the value of z^α is located on the circle $|w| = |z|^\alpha$, hence

(1) $z^\alpha = z^n$ is single valued if $\alpha = a = n$ is an integer;

(2) when $\alpha = a = p/q$ is an irreducible fraction, and $0 < p < q$, p, q are integers, then

$$z^\alpha = e^{\frac{p}{q}\log|z|}e^{\mathrm{i}\frac{p}{q}(\arg z+2k\pi)} = |z|^{\frac{p}{q}}e^{\mathrm{i}\frac{p}{q}\arg z}e^{\mathrm{i}\frac{p}{q}2k\pi}.$$

The numbers $\frac{p}{q}2k\pi$, $k = 0, \pm 1, \cdots$, are not congruent each other with 2π provided $k = 0, 1, 2, \cdots, q - 1$. Any other value of k, $\frac{p}{q}2k\pi$ is congruent with one of them. Thus z^α has q different values only.

(3) z^α is an infinity many valued function if $\alpha = a$ is an irrational number.

§ 1.6   Complex Series

Finally, there are several results in the theory of series in calculus which can easily extend to the field of complex numbers.

For example: a sequence of functions $\{f_n(z)\}$ on a set $E \subset \mathbb{C}$ converges uniformly to $f(z)$ means that for any given $\varepsilon > 0$, there exists a positive integer n_0, which depends on ε and is independent of z, such that

$$|f_n(z) - f(z)| < \varepsilon$$

hlods for all $n \geq n_0$ and all $z \in E$.

Just like in calculus, we may prove that the limit function of a uniformly convergence sequence of continuous functions is continuous.

Cauchy criterion   A sequence of functions $\{f_n(z)\}$ on the set $E \subset \mathbb{C}$ converges uniformly if and only if for any given $\varepsilon > 0$, there exists a positive integer n_0, which depends on ε and is independent of z, such that

$$|f_m(z) - f_n(z)| < \varepsilon$$

hlods for all $m, n \geq n_0$ and all $z \in E$.

Weierstrass M-test   Suppose a series of functions

$$f_1(z) + f_2(z) + \cdots + f_n(z) + \cdots$$

on a set $E \subseteq \mathbb{C}$ is defined, and $a_1 + a_2 + a_3 + \cdots$ is a series of positive terms. If, there exists a positive integer n_0 and a constant $M > 0$, such that $|f_n(z)| \leq M a_n$ hlods true for all $z \in E$ when $n > n_0$, then $\sum\limits_{n=1}^{\infty} f_n(z)$ converges uniformly on E if $\sum\limits_{n=1}^{\infty} a_n$ converges.

In particular, we consider the power series

$$a_0 + a_1 z + a_2 z^2 + \cdots + a_n z^n + \cdots . \tag{6.1}$$

We have the following theorem.

Theorem 3 (Abel theorem)   For power series (6.1), there exists a number R, $0 \leq R \leq \infty$, which is the radius of convergence of the series, and has the following properties:

(1) The series converges absolutely for every z in $|z| < R$. The series converges uniformly on $|z| \leq \rho$ if $0 \leq \rho < R$.

(2) The terms of the series is unbounded, and the series diverges when z in $|z| > R$.

(3) The sum of the series is a holomorphic function if $|z| < R$. Its derivative can obtain by differentiating the series term by term, and the radius of convergence of the series after differentiation is the same as the radius of convergence of the original series.

The disc $|z| \leq R$ is the convergence circle. We do not know the series converges or not on the circumference of the disc. The value R is

$$R = \frac{1}{\limsup\limits_{n \to \infty} \sqrt[n]{|a_n|}}. \tag{6.2}$$

This is the Hadamard formula of radius of convergence.

If $R = 0$, the series diverges everywhere except $z = 0$; If $R = \infty$ the series converges everywhere.

Now we prove Theorem 3.

Proof If $|z| < R$, we choose ρ, $|z| < \rho < R$, then $\frac{1}{\rho} > \frac{1}{R}$. By (6.2), there exists an integer n_0, so that $|a_n|^{\frac{1}{n}} < \frac{1}{\rho}$ when $n \geq n_0$, thus $|a_n| < \frac{1}{\rho^n}$ and $|a_n z^n| < (\frac{|z|}{\rho})^n$ when $n \geq n_0$. The series $\sum\limits_{n=0}^{\infty} (\frac{|z|}{\rho})^n$ converges when $|z| < \rho$, then $\sum\limits_{n=0}^{\infty} a_n z^n$ converges absolutely by Weierstrass M-test. In purpose to prove the series converges uniformly on $|z| \leq \rho$ $(< R)$, we choose ρ', n_1 so that $\rho < \rho' < R$, and $|a_n z^n| < (\frac{\rho}{\rho'})^n$ when $n \geq n_1$. The series converges uniformly by Weierstrass M-test again.

If $|z| > R$, we choose ρ, $R < \rho < |z|$, then $\frac{1}{\rho} < \frac{1}{R}$. There exists a positive integer n_2 and a subsequence $\{m_i\}$, so that $|a_{m_i}|^{\frac{1}{m_i}} > \frac{1}{\rho}$, that is $|a_{m_i}| > \frac{1}{\rho^{m_i}}$ when $m_i \geq n_2$. Hence there are infinity many n for $|a_n z^n| > (\frac{|z|}{\rho})^n$. It implies that the series is unbounded.

The series $\sum\limits_{n=1}^{\infty} n a_n z^n$ and the series $\sum\limits_{n=1}^{\infty} a_n z^n$ have the same radius of convergence because $\lim\limits_{n \to \infty} \sqrt[n]{n} = 1$.

For $|z| < R$,

$$f(z) = \sum_{n=0}^{\infty} a_n z^n = S_n(z) + R_n(z),$$

where

$$S_n(t) = a_0 + a_1 z + \cdots + a_{n-1} z^{n-1}, \qquad R_n(z) = \sum_{k=n}^{\infty} a_k z^k.$$

Let

$$f_1(z) = \sum_{n=1}^{\infty} n a_n z^{n-1} = \lim_{n \to \infty} S'_n(z).$$

We will prove that $f_1(z) = f'(z)$.

Choose ρ, $0 < \rho < R$, and fix a point z_0, $|z_0| < \rho$, we have

$$\frac{f(z) - f(z_0)}{z - z_0} - f_1(z) = \left(\frac{S_n(z) - S_n(z_0)}{z - z_0} - S'_n(z_0) \right) + \left(S'_n(z_0) - f_1(z) \right)$$
$$+ \left(\frac{R_n(z) - R_n(z_0)}{z - z_0} \right).$$

If $z \neq z_0$, and $|z| < \rho < R$, the last term of the right hand side of the previous equality is

$$\sum_{k=n}^{\infty} a_k (z^{k-1} + z^{k-2} z_0 + \cdots + z z_0^{k-2} + z_0^{k-1}),$$

hence

$$\left| \frac{R_n(z) - R_n(z_0)}{z - z_0} \right| \leq \sum_{k=n}^{\infty} k |a_k| \rho^{k-1}.$$

The right hand side is the remainder term of a convergence series, there exists a positive integer n_0, so that

$$\left| \frac{R_n(z) - R_n(z_0)}{z - z_0} \right| < \frac{\varepsilon}{3}$$

when $n \geq n_0$. Since $f_1(z) = \lim_{n \to \infty} S'_n(z)$, there exists an integer n_4, $|S'_n(z_0) - f_1(z)| < \varepsilon/3$ when $n \geq n_4$. Fix n, $n > n_3$, $n > n_4$, by the definition of derivative, there exists $\delta > 0$, so that

$$\left| \frac{S_n(z) - S_n(z_0)}{z - z_0} - S'_n(z_0) \right| < \frac{\varepsilon}{3}$$

when $0 < |z - z_0| < \delta$.

Combiniing above results, we have

$$\left| \frac{f(z) - f(z_0)}{z - z_0} - f_1(z) \right| < \varepsilon$$

when $0 < |z - z_0| < \delta$. It proved $f'(z) = f_1(z)$.

Theorem 3 have proved.

Repeating this procedure several times, we have

$$f^{(k)}(z) = k!a_k + \frac{(k+1)!}{1!}a_{k+1}z + \frac{(k+2)!}{2!}a_{k+2}z^2 + \cdots$$

holds true for any positive integer k. Hence $a_k = \frac{f^{(k)}(0)}{k!}$, the power series can be expressed as

$$f(z) = f(0) + f'(0)z + \frac{f''(0)}{2!}z^2 + \cdots + \frac{f^{(n)}(0)}{n!}z^n + \cdots .$$

This is the Tayolr-Maclaurin series. It was proved under the assumption that $f(z)$ has a power series. The expansion of $f(z)$ is unique if it exists. In next chapter, we will prove that every holomorphic function has a Taylor expansion.

By the property (4) of exponebtial functions, $(e^z)' = e^z$, the Taylor series of e^z is

$$e^z = 1 + \frac{z}{1!} + \frac{z^2}{2!} + \cdots + \frac{z^n}{n!} + \cdots .$$

By the definition (5.3), the Taylor series of $\cos z$ and $\sin z$ are

$$\cos z = 1 - \frac{z^2}{2!} + \frac{z^4}{4!} - \frac{z^6}{6!} + \cdots ,$$

$$\sin z = z - \frac{z^3}{3!} + \frac{z^5}{5!} - \frac{z^7}{7!} + \cdots .$$

These three series converge on the complex plane by Hadamard formula.

Finally, we state two results about the series of the functions.

Theorem 4   (1) If functions in the sequence $\{f_n(z)\}$ $(n = 1, 2, \cdots)$ are continuous on the set A, and $\sum\limits_{n=1}^{\infty} f_n(z)$ converges uniformly to $f(z)$ on A, then $f(z)$ is continuous on A.

(2) If functions in the sequence $\{f_n(z)\}$ $(n = 1, 2, \cdots)$ are continuous on a rectifiable curve γ, and $\sum\limits_{n=1}^{\infty} f_n(z)$ converges uniformly to $f(z)$ on γ, then

$$\int_{\gamma} f(z)\,dz = \sum_{n=1}^{\infty} \int_{\gamma} f_n(z)\,dz.$$

These two results are the same as the corresponding theorems of the series of functions in calculus, and the proofs are the same, we omit the details of the proofs.

EXERCISES I

1. Use formula (1.9) to verify the Newton-Leibniz formula, the Green formula, the Stokes formula and the Gauss formula.

2. (1) Find the modulus and principal argument of each of the following complex numbers:

(i) $2\,i$; (ii) $1-i$; (iii) $3+4\,i$; (iv) $-5+12\,i$.

(2) Find the value of each of the following complex numbers.

(i) $(1+3\,i)^3$; (ii) $\frac{10}{4-3\,i}$; (iii) $\frac{2-3\,i}{4+i}$; (iv) $(1+i)^n+(1-i)^n$ where n is a positive integer.

(3) Find the absolute value of each of the following complex numbers:

(i) $-3\,i\,(2-i)(3+2\,i)(1+i)$; (ii) $\dfrac{(4-3\,i)(2-i)}{(1+i)(1+3\,i)}$.

3. Evaluate $(5-i)^4(1+i)$, and then to prove

$$\frac{\pi}{4}=4\arctan\frac{1}{5}-\arctan\frac{1}{239}.$$

4. Suppose $z=x+i\,y$, x,y are real numbers, to find the real part and imaginary part of each of the following complex numbers

(1) $\dfrac{1}{z}$; (2) z^2; (3) $\dfrac{1+z}{1-z}$; (4) $\dfrac{z}{z^2+1}$.

5. Solve the quadratic equation

$$z^2+(\alpha+i\,\beta)a+\gamma+i\,\delta=0$$

where $\alpha,\beta,\gamma,\delta$ are real numbers.

6. Suppose $|z|=r>0$. Show that

$$\operatorname{Re}z=\frac{1}{2}\left(z+\frac{r^2}{z}\right),\qquad \operatorname{Im}z=\frac{1}{2\,i}\left(z-\frac{r^2}{z}\right).$$

7. Show that : (1) $\left|\frac{a-b}{1-\bar{a}b}\right|<1$ if $|a|<1$, $|b|<1$;

(2) $\left|\dfrac{a-b}{1-\bar{a}b}\right|=1$ if $|a|=1$ or $|b|=1$. Consider the case $|a|=1$ and $|b|=1$.

8. Prove the Lagrange equality in complex form

$$\left|\sum_{i=1}^{n} a_i b_i\right|^2 = \sum_{i=1}^{n} |a_i|^2 \sum_{i=1}^{n} |b_i|^2 - \sum_{1 \le i < j \le n} |a_i \bar{b}_j - a_j \bar{b}_i|^2,$$

and then derive the Cauchy inequality

$$\left|\sum_{i=1}^{n} a_i b_i\right|^2 \le \sum_{i=1}^{n} |a_i|^2 \sum_{i=1}^{n} |b_i|^2$$

from the above equality. Show that the equality holds in above inequality if and only if the ratios of a_k and $\bar{b}_k$ are the same for $k = 1, \cdots, n$.

9. Show that: a_1, a_2, a_3 are the vertices of an equilateral triangle if and only if $a_1^2 + a_2^2 + a_3^2 = a_1 a_2 + a_2 a_3 + a_3 a_1$.

10. Show that the complex numbers α, β, γ are collinear if and only if

$$\begin{bmatrix} \alpha & \bar{\alpha} & 1 \\ \beta & \bar{\beta} & 1 \\ \gamma & \bar{\gamma} & 1 \end{bmatrix} = 0.$$

11. Find the corresponding points on Riemann sphere S^2 for $1 - i, 4 + 3i$ in the complex plane.

12. Suppose z_1, z_2 are two points on the complex plane $\mathbb{C}$. Show that their corresponding points on Riemann sphere S^2 are two terminal points of a diameter if and only if $z_1 \bar{z}_2 = -1$.

13. Show that the equation of a circle is $A|z|^2 + Bz + \overline{B}\bar{z} + C = 0$, where A, C are real numbers, and $|B|^2 > AC$. Moreover, the corresponding image of this circle on the Riemann sphere S^2 is a great circle if and only if $A + C = 0$.

14. Suppose z_1, z_2 are two points on $\mathbb{C}$, Z_1, Z_2 are their corresponding points on Riemann sphere S^2. and $d(z_1, z_2)$ denote the spherical distance between Z_1 and Z_2. Show that

$$d(z_1, z_2) = \frac{2|z_1 - z_2|}{\sqrt{(1 + |z_1|^2)(1 + |z_2|^2)}}, \qquad \text{where } z_1, z_2 \in \mathbb{C};$$

$$d(z_1, \infty) = \frac{2}{\sqrt{1 + |z_1|^2}}, \qquad \text{where } z_1 \in \mathbb{C}.$$

15. Explain the geometrical meaning of each of the following relations:

(i) $\left|\dfrac{z - z_1}{z - z_2}\right| \le 1$, where z_1 and z_2 are two fixed points, and $z_1 \ne z_2$;

(ii) $\mathrm{Re}\,\dfrac{z - z_1}{z - z_2} = 0$, where z_1 and z_2 are two fixed points, and $z_1 \ne z_2$;

(iii) $0 < \arg\dfrac{z + \mathrm{i}}{z - \mathrm{i}} < \dfrac{\pi}{4}$;

(iv) $|z + c| + |z - c| \le 2a$, where $a > 0$ and $|c| < a$.

16. Prove the Heine-Borel theorem and Bolzano-Weierstrass theorem in the complex plane.

17. Show that: a sequence $z_n \in \mathbb{C}$ $(n = 1, 2, \cdots)$ converges to a point $z_0 \in \mathbb{C}$ if and only if the sequences $\mathrm{Re}\,z_n$, and $\mathrm{Im}\,z_n$ $(n = 1, 2, \cdots)$ converge to $\mathrm{Re}\,z_0$ and $\mathrm{Im}\,z_0$ respectively.

18. Show that: (1) if $\lim\limits_{n\to\infty} z_n = a$, $\lim\limits_{n\to\infty} z'_n = b$, then

$$\lim_{n\to\infty} \frac{1}{n} \sum_{k=1}^{n} z_k z'_{n-k} = ab.$$

(2) then derive that

$$\lim_{n\to\infty} \frac{1}{n}(z_1 + \cdots + z_n) = A$$

if $\lim\limits_{n\to\infty} z_n = A$.

19. Consider the differentiability of each of the following functions:

(i) $f(z) = |z|$;   (ii) $f(z) = \bar{z}$;   (iii) $f(z) = \mathrm{Re}\,z$.

20. Show that $g(f(z))$ is a holomorphic function if $\zeta = g(w)$ and $w = f(z)$ are holomorphic functions.

21. Show that:

(1) if $f(z)$ is holomorphic on a domain D, and $f'(z)$ identically equals zero on D, then $f(z)$ is a constant on D.

(2) if $f(z)$ is holomorphic on a domain D, and satisfies one of the following conditions:

(i) $\mathrm{Re}\,f(z)$ is a constant on D;   (ii) $\mathrm{Im}\,f(z)$ is a constant on D;

(iii) $|f(z)|$ is a constant on D;   (iv) $\arg f(z)$ is a constant on D;

then $f(z)$ is a constant on D.

22. Show that if $f(z) = u + \mathrm{i}v$ is a holomorphic function, and $f'(z) \ne 0$, then the curves $u(x, y) = c_1$ and $v(x, y) = c_2$ are orthogonal where c_1, c_2 are constants.

23. Show that:

(1) Under the polar coordinates, if

$$f(z) = u(r,\theta) + i\,v(r,\theta), \qquad z = r(\cos\theta + i\,\sin\theta),$$

then the C-R equations are

$$u_r = \frac{1}{r}\,v_\theta, \qquad v_r = -\frac{1}{r}\,u_\theta,$$

and

$$f'(z) = \frac{r}{z}\,(u_r + i\,v_r).$$

(2) If we denote $f(z) = R(\cos\varphi + i\,\sin\varphi)$, then the C-R equations are

$$\frac{\partial R}{\partial r} = -\frac{R}{r}\frac{\partial\varphi}{\partial\theta}, \qquad \frac{\partial R}{\partial\theta} = -Rr\frac{\partial\varphi}{\partial r}.$$

24. Show that if $f(z) = u(x,y) + i\,v(x,y)$ $(z = x + i\,y)$ is holomorphic on a domain D, then the Jacobian is

$$J = \begin{vmatrix} u_x & u_y \\ v_x & v_y \end{vmatrix} = |f'(z)|^2.$$

Explain its geometric meaning.

25. Evaluate each of the following line integrals.

(i) $\displaystyle\int_\gamma x\,dz$, where $z = x + i\,y$, γ is a directional segment from 0 to $1 + i$;

(ii) $\displaystyle\int_\gamma |z - 1|\,|dz|$, where $\gamma(t) = e^{i\,t}$, $0 \le t \le 2\pi$;

(iii) $\displaystyle\int_\gamma \frac{1}{z - a}\,dz$, where $\gamma(t) = a + Re^{i\,t}$, $0 \le t \le 2\pi$, a is a complex number.

26. (i) Find the real part and imaginary part of $\cos(x + i\,y)$ and $\sin(x + i\,y)$, where x, y are real numbers;

(ii) Show that:

$$\sin i\,z = i\,\mathrm{sh}\,z, \qquad \cos i\,z = \mathrm{ch}\,z, \qquad (\sin z)' = \cos z,$$
$$\cos(z_1 + z_2) = \cos z_1 \cos z_2 - \sin z_1 \sin z_2.$$

27. Evaluate the value of each of the following expressions: $\sin i, \cos(2 + i), \tan(1 + i), 2^i, i^i, (-1)^{2i}, \log(2 - 3\,i), \arccos\frac{1}{4}(3 + i)$.

28. (i) Evaluate the value e^z when $z = \frac{\pi i}{2}, -\frac{3}{2}\pi i$;

(ii) Solve z from the equation $e^z = i$.

29. Find the real part and imaginary part of z^z where $z = x + iy$.

30. Show that the roots of the equation $z^n = a$ are the vertices of a equilateral polygon.

31. Prove that (1) Cauchy criterion, and (2) Weierstrass M-test hold in the complex number field.

32. Find the radius of convergence of the series $\sum\limits_{n=1}^{\infty} a_n z^n$ if

(i) $a_n = n^{\frac{1}{n}}$; (ii) $a_n = n^{\log n}$; (iii) $a_n = \dfrac{n!}{n^n}$; (iv) $a_n = n^n$.

33. Prove Theorem 4.

34. Suppose $f(z)$ is holomorphic on $\mathbb{C}$, show that

(i) if $f'(z) = f(z)$ holds for every $z \in \mathbb{C}$, then $f(z) = e^z$;

(ii) if $f(z + w) = f(z)f(w)$ holds for every $z \in \mathbb{C}$ and every $w \in \mathbb{C}$, and $f'(0) = 1$, then $f(z) = e^z$.

35. Suppose f is holomorphic on $\mathbb{C}\backslash(-\infty, 0]$ and $f(1) = 0$, where $\mathbb{C}\backslash(-\infty, 0]$ means $\mathbb{C}$ minus the segment $(-\infty, 0]$ (In the later of this book, we will use this notation again, and do not explain its meaning). Show that

(i) if $f'(z) = e^{-f(z)}$ holds for every $z \in \mathbb{C}\backslash(-\infty, 0]$, then

$$f(z) = \log z;$$

(ii) if $f(zw) = f(z) + f(w)$ holds for every $z \in \mathbb{C}\backslash(-\infty, 0]$ and every $w \in \mathbb{C}\backslash(-\infty, 0]$, and $f'(1) = 1$, then

$$f(z) = \log z.$$

36. Show that: if $\varphi(z) = \frac{1}{2}(z + \frac{1}{z})$, then each of the following four domains is the univalent domain of φ:

(i) upper half plane $\{z \in \mathbb{C} \mid \operatorname{Im} z > 0\}$;

(ii) lower half plane $\{z \in \mathbb{C} \mid \operatorname{Im} z < 0\}$;

(iii) punctured unit disk $D(0, 1)\backslash\{0\}$, where $D(0, 1)$ is the unit disc centered at origin with radius 1;

(iv) exterior part of the unit disc $\{z \in \mathbb{C} \mid |z| > 1\}$.

37. Find the images of these four domains in the previous exercise by $\varphi(z) = \frac{1}{2}(z + \frac{1}{z})$.

38. Show that each of the following three domains is the univalent domain of $\cos z$ and $\sin z$:

(i) Strip $\{z \in \mathbb{C} \mid \theta_0 < \operatorname{Re} z < \theta_0 + \pi\}$ where θ_0 is a fix real number;

(ii) Semi-strip $\{z \in \mathbb{C} \mid \theta_0 < \operatorname{Re} z < \theta_0 + 2\pi, \ \operatorname{Im} z > 0\}$, where θ_0 is a fix real number;

(iii) Semi-strip $\{z \in \mathbb{C} \mid \theta_0 < \operatorname{Re} z < \theta_0 + 2\pi, \ \operatorname{Im} z < 0\}$ where θ_0 is a fix real number.

39. Show that $w = \cos z$ is an univalent function of the semi-strip $\{z \in \mathbb{C} \mid 0 < \operatorname{Re} z < 2\pi, \ \operatorname{Im} z > 0\}$ onto $\mathbb{C} \backslash [-1, +\infty)$.

40. Show that $w = \sin z$ is an univalent function of the semi-strip $\{z \in \mathbb{C} \mid -\frac{\pi}{2} < \operatorname{Re} z < \frac{\pi}{2}, \ \operatorname{Im} z > 0\}$ onto the upper half plane.

41. Show that $D(0,1)$ is the univalent domain of $f(z) = \frac{z}{(1-z)^2}$. Find the image $f(D(0,1))$.

42. When z rotates along the circle $\{z \in \mathbb{C} \mid |z| = 2\}$ once in counter-clockwise, to evaluate the increment of the argument of each of the following functions:

(i) $(z-1)^{\frac{1}{2}}$;   (ii) $(1+z^4)^{\frac{1}{3}}$;   (iii) $(z^2+2z-3)^{\frac{1}{4}}$;

(iv) $\left(\dfrac{z-1}{z+1}\right)^{\frac{1}{2}}$;   (v) $\left(\dfrac{z^2-1}{z^2+5}\right)^{\frac{1}{7}}$.

43. Show that if $f(z)$ is a holomorphic univalent function on a domain U, then the area of $f(U)$ is

$$\iint_U |f'(z)|^2 \mathrm{d}x\,\mathrm{d}y.$$

where $z = x + \mathrm{i}y$.

44. Show that if $\{a_n\}$ and $\{b_n\}$ satisfy the following conditions:

(i) $\{S_n\}$ is bounded, where $S_n = \sum\limits_{k=1}^{n} a_k$;

(ii) $\lim\limits_{n \to \infty} b_n = 0$;

(iii) $\sum\limits_{n=1}^{\infty} |b_n - b_{n+1}| < +\infty$,

then the series $\sum\limits_{n=1}^{\infty} a_n b_n$ converges. Explain that it is the extension of the Dirichlet criterion and the Abel criterion of the series of real number terms.

45. Show that if $\sum\limits_{n=1}^{\infty} a_n$ is a series of complex number terms, and $\varlimsup\limits_{n \to \infty} \sqrt[n]{|a_n|} = q$, then

(i) $\sum\limits_{n=1}^{\infty} a_n$ converges absolutely if $q < 1$;

(ii) $\sum\limits_{n=1}^{\infty} a_n$ diverges if $q > 1$.

46. Show that if $a_n \in \mathbb{C}\backslash\{0\}$, $n = 1, 2, \cdots$, and $\varliminf\limits_{n\to\infty} |\frac{a_{n+1}}{a_n}| = q$, then $\sum\limits_{n=1}^{\infty} a_n$ converges alsolutely when $q < 1$. Does the series $\sum\limits_{n=1}^{\infty} a_n$ converge or diverge when $q > 1$?

47. Show that if $a_n \in \mathbb{C}\backslash\{0\}$ $(n = 1, 2, \cdots)$, $\varlimsup\limits_{n\to\infty} |\frac{a_{n+1}}{a_n}| = 1$ and $\varliminf\limits_{n\to\infty} n(|\frac{a_{n+1}}{a_n}| - 1) < -1$, then $\sum\limits_{n=1}^{\infty} a_n$ converges absolutely (Reabe criterion).

48. Show that if $\{a_n\}$ $(n = 0, 1, 2, \cdots)$ is a positive monotonic sequence and it approaches to zero, the radius of convergence of $\sum\limits_{n=0}^{\infty} a_n z^n$ is R then

(i) $R \geq 1$;

(ii) $\sum\limits_{n=0}^{\infty} a_n z^n$ converges everywhere on $\partial D(0, R)\backslash\{R\}$, where $\partial D(0, R)$ means the boundary of the disk $D(0, R)$, the circle centered at origin with radius R.

49. Show that the power series $\sum\limits_{n=1}^{\infty} a_n(z - z_0)^n$ converges uniformly on its convergence disk D if and only if it converges uniformly on $\overline{D}$, where $z_0 \in \mathbb{C}$ is a fixed point, and $\overline{D}$ is the closure of D.

50. Show that if the radius of convergence of the power series $\sum\limits_{n=0}^{\infty} a_n z^n = f(z)$ is 1, $z_0 \in \partial D(0, 1)$, then $\sum\limits_{n=0}^{\infty} a_n z_0^n$ converges to $\lim\limits_{r\to 1} f(rz_0)$ if $\lim\limits_{n\to\infty} na_n = 0$ and $\lim\limits_{r\to 1} f(rz_0)$ exists.

CHAPTER II
CAUCHY INTEGRAL THEOREM AND
CAUCHY INTEGRAL FORMULA

§ 2.1   Cauchy-Green Formula (Pompeiu Formula)

The theory of Cauchy integral is one of the three main parts in the theory of functions of one complex variable. The theory of functions of one complex variable becomes an individual branch of mathematics only when the theory of Cauchy integral was established. After that, a series of interesting results was obtained which was essentially different from the calculus of real variable.

We start from the Cauchy-Green formula which is the direct consequence of Theorem 2 (Green formula in complex form) of last chapter.

Theorem 1 (Cauchy-Green formula, Pompeiu formula)   Let $U \subset \mathbb{C}$ be a bounded domain with C^1 boundary, i.e., the boundary is a smooth curve. Let $f(z) = u(x,y) + iv(x,y) \in C^1(\overline{U})$, that means $u(x,y), v(x,y)$ have the first order continuous partial derivatives, then for any $z \in U$,

$$
\begin{aligned}
f(z) &= \frac{1}{2\pi i} \int_{\partial U} \frac{f(\zeta)}{\zeta - z} \, d\zeta - \frac{1}{2\pi i} \iint_U \frac{\partial f(\zeta)}{\partial \overline{\zeta}} \cdot \frac{d\overline{\zeta} \wedge d\zeta}{\zeta - z} \\
&= \frac{1}{2\pi i} \int_{\partial U} \frac{f(\zeta)}{\zeta - z} \, d\zeta - \frac{1}{\pi} \iint_U \frac{\partial f(\zeta)}{\partial \overline{\zeta}} \cdot \frac{dA(s)}{\zeta - z}.
\end{aligned}
\tag{1.1}
$$

Proof   Let $D(z, \varepsilon)$ be a disk with small radius ε, centered at z, and $D(z, \varepsilon) \subset U$. Let $U_{z,\varepsilon} = U \backslash D(z, \varepsilon)$. Consider the differential form

$$
\frac{f(\zeta) \, d\zeta}{\zeta - z}
$$

on $U_{z,\varepsilon}$. Then

$$
\int_{\partial U} \frac{f(\zeta) \, d\zeta}{\zeta - z} - \int_{\partial D_{z,\varepsilon}} \frac{f(\zeta) \, d\zeta}{\zeta - z} = \iint_{U_{z,\varepsilon}} d_\zeta \left(\frac{f(\zeta) \, d\zeta}{\zeta - z} \right)
$$

by the Theorem 2 (Green formula in complex form) of Chapter I. According to the definition of d_ζ, we have

$$\iint_{U_{z,\varepsilon}} d_\zeta \left(\frac{f(\zeta)\,d\zeta}{\zeta - z} \right) = \iint_{U_{z,\varepsilon}} (\partial + \overline{\partial}) \left(\frac{f(\zeta)d\zeta}{\zeta - z} \right)$$

$$= \iint_{U_{z,\varepsilon}} \partial \left(\frac{f(\zeta)\,d\zeta}{\zeta - z} \right) + \iint_{U_{z,\varepsilon}} \overline{\partial} \left(\frac{f(\zeta)\,d\zeta}{\zeta - z} \right).$$

We know that

$$\partial \left(\frac{f(\zeta)\,d\zeta}{\zeta - z} \right) = \frac{\partial}{\partial \zeta} \left(\frac{f(\zeta)}{\zeta - z} \right) d\zeta \wedge d\zeta = 0,$$

and

$$\overline{\partial} \left(\frac{f(\zeta)\,d\zeta}{\zeta - z} \right) = \frac{\partial f}{\partial \overline{\zeta}} \frac{d\overline{\zeta} \wedge d\zeta}{\zeta - z} + f \frac{\partial}{\partial \overline{\zeta}} \left(\frac{1}{\zeta - z} \right) d\overline{\zeta} \wedge d\zeta$$

$$= \frac{\partial f}{\partial \overline{\zeta}} \frac{d\overline{\zeta} \wedge d\zeta}{\zeta - z}$$

since $\frac{\partial}{\partial \overline{\zeta}} \left(\frac{1}{\zeta - z} \right) = 0$. Thus

$$\iint_{U_{z,\varepsilon}} d_\zeta \left(\frac{f(\zeta)\,d\zeta}{\zeta - z} \right) = \iint_{U_{z,\varepsilon}} \frac{\partial f}{\partial \overline{\zeta}} \frac{d\overline{\zeta} \wedge d\zeta}{\zeta - z}.$$

On the other hand,

$$\int_{\partial D(z,\varepsilon)} \frac{f(\zeta)\,d\zeta}{\zeta - z} = \int_{\partial D(z,\varepsilon)} \frac{f(\zeta) - f(z)}{\zeta - z}\,d\zeta + \int_{\partial D(z,\varepsilon)} \frac{f(z)\,d\zeta}{\zeta - z}.$$

By assumption $f(\zeta) \in C^1(\overline{U})$, there exists a constant C, such that

$$|f(\zeta) - f(z)| < C|\zeta - z|$$

holds for $\zeta \in \partial D(z, \varepsilon)$. Hence

$$\left| \int_{\partial D(z,\varepsilon)} \frac{f(\zeta) - f(z)}{\zeta - z}\,d\zeta \right| < C \int_{\partial D(z,\varepsilon)} \left| \frac{\zeta - z}{\zeta - z} \right| |d\zeta| = 2\pi\varepsilon C.$$

This implies $\int_{\partial D(z,\varepsilon)} \frac{f(\zeta) - f(z)}{\zeta - z}\,d\zeta \to 0$ when $\varepsilon \to 0$.

We may express $\zeta - z = \varepsilon e^{i\theta}$ $(0 \le \theta \le 2\pi)$ when $\zeta \in \partial D(z, \varepsilon)$, then

$$\int_{\partial D(z,\varepsilon)} \frac{f(z)\,\mathrm{d}\zeta}{\zeta - z} = f(z) \int_0^{2\pi} \frac{\varepsilon e^{i\theta} i\,\mathrm{d}\theta}{\varepsilon e^{i\theta}} = 2\pi\,\mathrm{i}\,f(z).$$

Thus

$$\int_{\partial U} \frac{f(\zeta)\,\mathrm{d}\zeta}{\zeta - z} - 2\pi\mathrm{i}f(z) = \iint\limits_{U_{z,\varepsilon}} \frac{\partial f}{\partial \overline{\zeta}} \frac{\mathrm{d}\overline{\zeta} \wedge \mathrm{d}\zeta}{\zeta - z} + O(\varepsilon)$$

where $O(\varepsilon)$ denotes a term which approaches to a constant when it divided by ε and let $\varepsilon \to 0$. We have (1.1) when we let $\varepsilon \to 0$ in the previous equality.

As some consequences, we have

Theorem 2 (Cauchy integral formula)   Let $U \subseteq \mathbb{C}$ be a bounded domain with C^1 boundary, and let $f(z)$ be a holomorphic function on U and $f(z) \in C^1(\overline{U})$, then for any $z \in U$,

$$f(z) = \frac{1}{2\pi\mathrm{i}} \int_{\partial U} \frac{f(\zeta)}{\zeta - z}\,\mathrm{d}\zeta. \tag{1.2}$$

Theorem 3 (Cauchy integral theorem)   Let $U \subseteq \mathbb{C}$ be a bounded domain with C^1 boundary, and let $F(z)$ be a holomorphic function on U and $F(z) \in C^1(\overline{U})$, then

$$\int_{\partial U} F(\zeta)\,\mathrm{d}\zeta = 0. \tag{1.3}$$

Proof   We may assume that U contains the origin. Let $f(z) = zF(z)$ in (1.2), then we have (1.3) when we let $z = 0$.

Thus the Cauchy integral formula implies the Cauchy integral theorem. Of course, we may prove Theorem 3 directly from the Cauchy-Green theorem. Conversely, the Cauchy integral theorem implies the Cauchy integral formula by the following process. Fix a point $z_0 \in U$, and consider $U_{z_0,\varepsilon}$ as we defined at Theorem 1. Let $F(z) = \frac{f(z)}{z - z_0}$, then we may prove Theorem 2 just like Theorem 1 we proved. Thus, Theorem 2 and Theorem 3 are equivalent, and these two equivalent theorems are the important key stones of the theory of functions of one complex variable.

Another important application of the Cauchy-Green formula is to solve one-dimension $\overline{\partial}$-equation. It will be used at Chapter III.

Let ψ be a continuous function, the closure of set of points for which $\psi \neq 0$ is the support of ψ, denoted by **supp ψ**.

Theorem 4 (Solution of one-dimension $\overline{\partial}$-equation)　Let $\psi(z) \in C^1(C)$ with a compact support, i.e., the support is compact. Let

$$u(z) = \frac{-1}{2\pi i} \iint_C \frac{\psi(\zeta)}{\zeta - z} \, d\overline{\zeta} \wedge d\zeta, \tag{1.4}$$

then $u(z) \in C^1(\mathbb{C})$ and $u(z)$ is the solution of $\overline{\partial}$-equation $\frac{\partial u(z)}{\partial \overline{z}} = \psi(z)$.

Proof　Fix $z \in \mathbb{C}$, let $\xi = \zeta - z$, then

$$u(z) = \frac{-1}{2\pi i} \iint_{\mathbb{C}} \frac{\psi(\xi + z)}{\xi} d\overline{\xi} \wedge d\xi.$$

$u(z)$ is a continuous function since $\frac{1}{\xi}$ is integrable on any compact set. If $h \in \mathbb{R}, h \neq 0$,

$$\frac{u(z+h) - u(z)}{h} = \frac{-1}{2\pi i} \iint_{\mathbb{C}} \frac{1}{\xi} \cdot \frac{\psi(\xi + z + h) - \psi(\xi + z)}{h} \, d\overline{\xi} \wedge d\xi.$$

Fix z and ξ, and let $h \to 0$. Since $\psi \in C^1(\mathbb{C})$ and ψ has a compact support,

$$\frac{\psi(\xi + z + h) - \psi(\xi + z)}{h} \longrightarrow \frac{\partial \psi(\xi + z)}{\partial \xi}$$

uniformly for ξ and z. Thus

$$\begin{aligned}
\frac{\partial u}{\partial x}(z) &= \lim_{h \to 0} \frac{1}{h}(u(z+h) - u(z)) \\
&= \frac{-1}{2\pi i} \iint_{\mathbb{C}} \frac{1}{\xi} \frac{\partial \psi}{\partial \alpha}(\xi + z) \, d\overline{\xi} \wedge d\xi \\
&= \frac{-1}{2\pi i} \iint_{\mathbb{C}} \frac{\partial \psi(\zeta)}{\partial \alpha} \frac{1}{\zeta - z} \, d\overline{\xi} \wedge d\xi
\end{aligned} \tag{1.5}$$

where $\zeta = \alpha + i\beta$, $\alpha, \beta \in \mathbb{R}$. Since $\frac{1}{|\xi|}$ is integrable on any compact set, and the limit is uniform for any compact set of z, thus $\frac{\partial u}{\partial x}$ is continuous. Similarly

$$\begin{aligned}
\frac{\partial u}{\partial y}(z) &= \frac{-1}{2\pi i} \iint_{\mathbb{C}} \frac{1}{\xi} \frac{\partial \psi}{\partial \beta}(\xi + z) \, d\overline{\xi} \wedge d\xi \\
&= \frac{-1}{2\pi i} \iint_{\mathbb{C}} \frac{\partial \psi(\zeta)}{\partial \beta} \frac{1}{\zeta - z} \, d\overline{\xi} \wedge d\xi
\end{aligned} \tag{1.6}$$

and $\frac{\partial u}{\partial y}$ is continuous. Thus $u \in C^1(\mathbb{C})$.

From (1.5) and (1.6), we have

$$\frac{\partial u}{\partial \overline{z}} = \frac{-1}{2\pi i} \iint_{\mathbb{C}} \frac{\partial \psi(\zeta)}{\partial \overline{\zeta}} \frac{1}{\zeta - z} \, d\overline{\xi} \wedge d\xi. \tag{1.7}$$

We know that $\psi(z)$ has a compact support supp ψ, then there exist $R > 0$, so that supp $\psi \subset D(0, R) = \{z \mid |z| < R\}$. We have

$$\frac{\partial u}{\partial \overline{z}} = \frac{-1}{2\pi i} \iint_{D(0, R+\varepsilon)} \frac{\partial \psi}{\partial \overline{\zeta}} \frac{1}{\zeta - z} \, d\overline{\zeta} \wedge d\zeta$$

by (1.7) where $\varepsilon > 0$. Using Cauchy-Green formula (Theorem 1), the right-hand side of the previous equality is equal to

$$\psi(z) - \frac{1}{2\pi i} \int_{\partial D(0, R+\varepsilon)} \frac{\psi(\zeta)}{\zeta - z} \, d\zeta.$$

Obviously, $\frac{1}{2\pi i} \int_{\partial D(0, R+\varepsilon)} \frac{\psi(\zeta)}{\zeta - z} \, d\zeta$ equals zero. We obtain

$$\frac{\partial u(z)}{\partial \overline{z}} = \psi(z).$$

We have proved the theorem.

Similarly, we may prove that: if $\psi(z) \in C^k(\mathbb{C})$ with a compact support, then $u(z) \in C^k(\mathbb{C})$ where $u(z)$ is the function defined by (1.4), and k is a positive integer or ∞. We may also prove that: if $\psi(z) \in C^k(\mathbb{C})$ and the support is the sum of finite or infinity many compact sets which are mutually non-intersecting, then $u(z) \in C^k(\mathbb{C})$ and $\frac{\partial u}{\partial \overline{z}} = \psi(z)$.

§ 2.2   Cauchy-Goursat Theorem

The original Cauchy integral formula and Cauchy integral theorem which were established by Cauchy are in the form of Theorem 2 and Theorem 3. After that, Goursat droped the condition $f(z) \in C^1(\overline{U})$, thus established the Cauchy-Goursat integral formula and theorem.

Theorem 2′ (Cauchy-Goursat integral formula)   Let $U \subseteq \mathbb{C}$ be a bounded domain, and let ∂U be a simple closed curve. If $f(z)$ is holomorphic on U, and continuous on $\overline{U}$, then for any $z \in U$,

$$f(z) = \frac{1}{2\pi i} \int_{\partial U} \frac{f(\zeta)}{\zeta - z} \, d\zeta. \tag{2.1}$$

Theorem 3' (Cauchy-Goursat integral theorem) Let $U \subseteq \mathbb{C}$ be a bounded domain, and let ∂U be a simple closed curve. If $f(z)$ is holomorphic on U, and continuous on $\overline{U}$, then for any $z \in U$,

$$\int_{\partial U} f(\zeta) \, d\zeta = 0. \tag{2.2}$$

Of course, Theorem 2' and Theorem 3' are equivalent. We prove Theorem 3' only by the classical method.

Lemma 1 Let $f(z)$ be a continuous function on the domain $G \subseteq \mathbb{C}$, and let Γ be any piece-wise smooth curve inside G, then for any small $\varepsilon > 0$, there exists a broken line P which is inside of G and inscribed in Γ, such that the inequality

$$\left| \int_{\Gamma} f(z) \, dz - \int_{P} f(z) \, dz \right| < \varepsilon$$

holds.

Proof Taking a closed subdomain $\overline{D} \subset G$, such that $\Gamma \subset \overline{D}$, then $f(z)$ is uniformly continuous on $\overline{D}$ since $f(z)$ is continuous on G. For any $\varepsilon > 0$, there exists $\delta = \delta(\varepsilon)$, such that $|f(z') - f(z'')| < \varepsilon/(2l)$ holds for any two points z', z'' whenever $|z' - z''| < \delta$, where l is the length of Γ. Dividing Γ into n arcs $s_0, s_1, \cdots, s_{n-1}$, the end points of s_j are z_j and z_{j+1}, $j = 0, 1, \cdots, n-1$, and the length of each arc is less than δ. Construct a broken line P containing the segments $l_0, l_1, \cdots, l_{n-1}$. The end points of segment l_j are z_j and z_{j+1}, $j = 0, 1, \cdots, n-1$.

The distance of any two points on each arc (and segment) is less than δ since the length of each arc (and segment) is less than δ. The integral $\int_{\Gamma} f(z) dz$ has an approximate value

$$s = f(z_0)\triangle z_0 + f(z_1)\triangle z_1 + \cdots + f(z_n)\triangle z_{n-1},$$

where $\triangle z_k = \int_{s_k} dz$. s can be expressed as

$$s = \int_{s_0} f(z_0) \, dz + \int_{s_1} f(z_1) \, dz + \cdots + \int_{s_{n-1}} f(z_{n-1}) \, dz.$$

The difference between $\int_\Gamma f(z)\mathrm{d}\,z$ and s is

$$\int_\Gamma f(z)\,\mathrm{d}\,z - s = \int_{s_0} (f(z) - f(z_0))\,\mathrm{d}\,z + \int_{s_1} (f(z) - f(z_1))\,\mathrm{d}\,z + \cdots$$
$$+ \int_{s_{n-1}} (f(z) - f(z_{n-1}))\,\mathrm{d}\,z.$$

Since $|f(z) - f(z_k)| < \varepsilon/(2l)$ when $z \in s_k$, we have

$$\left| \int_\Gamma f(z)\,\mathrm{d}\,z - s \right| < \varepsilon \frac{s_0}{2l} + \varepsilon \frac{s_1}{2l} + \cdots + \varepsilon \frac{s_{n-1}}{2l} = \frac{\varepsilon}{2}.$$

Of course, $\triangle z_k = \int_{l_k}\,\mathrm{d}\,z$, similarly, we have

$$\int_P f(z)\,\mathrm{d}z - s = \int_{l_0} (f(z) - f(z_0))\,\mathrm{d}z + \int_{l_1} (f(z) - f(z_1))\,\mathrm{d}z + \cdots$$
$$+ \int_{l_{n-1}} (f(z) - f(z_{n-1}))\,\mathrm{d}z.$$

Hence

$$\left| \int_P f(z)\,\mathrm{d}z - s \right| < \frac{\varepsilon l_0}{2l} + \frac{\varepsilon l_1}{2l} + \cdots + \frac{\varepsilon l_{n-1}}{2l} = \frac{\varepsilon(l_0 + l_1 + \cdots + l_{n-1})}{2l} < \frac{\varepsilon}{2}.$$

Thus

$$\left| \int_\Gamma f(z)\,\mathrm{d}z - \int_P f(z)\,\mathrm{d}z \right| \le \left| \int_\Gamma f(z)\,\mathrm{d}z - s \right| + \left| s - \int_P f(z)\,\mathrm{d}z \right|$$
$$\le \frac{\varepsilon}{2} + \frac{\varepsilon}{2} = \varepsilon.$$

We have proved Lemma 1.

Lemma 2   Let $f(z)$ be a holomorphic function on a simple connected domain $G \subseteq \mathbb{C}$, and let Γ be any piecewise smooth closed curve inside G, then

$$\int_\Gamma f(\zeta)\,\mathrm{d}\,\zeta = 0.$$

Proof   From Lemma 1, for any $\varepsilon > 0$, and any piecewise smooth closed curve Γ, there exists an inscribed broken line P and

$$\left| \int_\Gamma f(z)\,\mathrm{d}z - \int_P f(z)\,\mathrm{d}z \right| < \varepsilon$$

holds.

If Lemma 2 holds for any closed broken line P, i.e., $\int_P f(z)\mathrm{d}z = 0$, then Lemma 2 holds for any piecewise smooth closed curve.

For any closed broken line, we may decompose it as the sum of some triangles by adding some line segments. The values of integrations on these adding line segments cancel each other, thus the value of integration on the polygon is the same as the sum of the value of integrations on these triangles.

Thus Lemma 2 holds if it is true for any triangle.

Now we are going to prove Lemma 2 holds for any triangle.

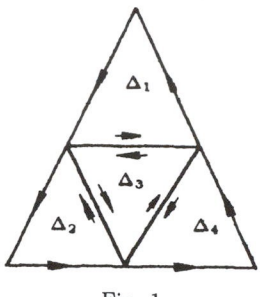

Fig. 1

Let $\triangle$ be a triangle inside G, and let M be the absolute value of the integration of $f(z)$ on $\triangle$,

$$M = \left| \int_{\triangle} f(z)\,\mathrm{d}z \right|.$$

We need to prove $M = 0$. Taking the middle point of three sides of the triangle, and joining these points, we divide the triangle into four small triangles which are congruent to each other. We denote the boundies of these four triangles by $\triangle_1, \triangle_2, \triangle_3$ and $\triangle_4$. Thus

$$\int_{\triangle} f(z)\,\mathrm{d}z = \left(\int_{\triangle_1} + \int_{\triangle_2} + \int_{\triangle_3} + \int_{\triangle_4} \right) f(z)\,\mathrm{d}z.$$

At least one of $\triangle_k$ ($k = 1,2,3,4$), has the property $|\int_{\triangle_k} f(z)\,\mathrm{d}z| \geq \frac{M}{4}$. We may assume it is $\triangle_1 = \triangle^{(1)}$. That is $|\int_{\triangle^{(1)}} f(z)\,\mathrm{d}z| \geq \frac{M}{4}$. Dividing $\triangle^{(1)}$ into four triangles by the same method, there exists a $\triangle^{(2)}$ at least, such that $|\int_{\triangle^{(2)}} f(z)\,\mathrm{d}z| \geq \frac{M}{4^2}$. We may repeat this process to get a sequence of triangles

$\triangle = \triangle^{(0)}$, $\triangle_1 = \triangle^{(1)}$, $\triangle^{(2)}, \cdots, \triangle^{(n)}, \cdots$. One includes the next one, and

$$\left| \int_{\triangle^{(n)}} f(z)\,dz \right| \geq \frac{M}{4^n}, \qquad n = 0, 1, 2, \cdots. \tag{2.3}$$

Let L denote the length of the circumference of $\triangle$, then the lengths of the circumferences of $\triangle^{(1)}, \triangle^{(2)}, \cdots, \triangle^{(n)}, \cdots$ are $\frac{L}{2}, \frac{L}{2^2}, \cdots, \frac{L}{2^n}, \cdots$. It approaches to zero when $n \to \infty$. There exists a point z_0 which belongs to all $\triangle^{(n)}$ ($n = 0, 1, 2, \cdots$). For any $\varepsilon > 0$, there exists $\delta = \delta(\varepsilon)$ such that

$$\left| \frac{f(z) - f(z_0)}{z - z_0} - f'(z_0) \right| < \varepsilon$$

holds when $|z - z_0| < \delta$. That is

$$|f(z) - f(z_0) - f'(z_0)(z - z_0)| < \varepsilon |z - z_0|$$

holds when $|z - z_0| < \delta$. When n is sufficiently large, $\triangle^{(n)}$ is inside $D(z_0, \varepsilon)$. Obviously, $\int_{\triangle^{(n)}}\,dz = 0$ and $\int_{\triangle^{(n)}} z\,dz = 0$, we have

$$\int_{\triangle^{(n)}} f(z)\,dz = \int_{\triangle^{(n)}} \left(f(z) - f(z_0) - (z - z_0)f'(z_0) \right)\,dz.$$

Hence

$$\left| \int_{\triangle^{(n)}} f(z)\,dz \right| < \int_{\triangle^{(n)}} \varepsilon |z - z_0|\,|dz|.$$

Since $|z - z_0|$ is the distance of a point z on $\triangle^{(n)}$ to the inner point z_0, we have

$$|z - z_0| < \frac{L}{2^n}.$$

It follows

$$\left| \int_{\triangle^{(n)}} f(z)\,dz \right| < \varepsilon \frac{L}{2^n} \cdot \frac{L}{2^n} = \varepsilon \frac{L^2}{4^n}. \tag{2.4}$$

From (2.3) and (2.4), we have $M < \varepsilon L^2$ for any $\varepsilon > 0$. It implies $M = 0$. The lemma have proved.

Proof of Theorem 3′   We prove Theorem 3′ for some special domain U at first.

Let U be a domain bounded by $x = a, x = b$ ($a < b$), and two rectifiable continuous curves:

$$MN: \quad y = \varphi(x), \qquad a \leq x \leq b,$$
$$PQ: \quad y = \psi(x), \qquad a \leq x \leq b,$$

where $\varphi(x) < \psi(x)$ $(a < x < b)$.

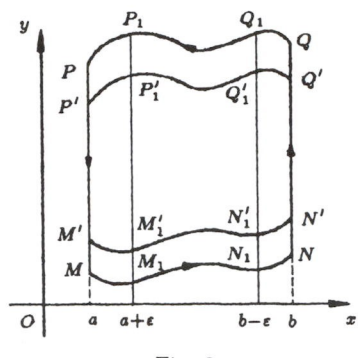

Fig. 2

If $f(z)$ is holomorphic on U and continuous on $\overline{U}$, we need to prove

$$\int_{MNQPM} f(z)\,dz = 0. \tag{2.5}$$

Construct two straight lines $x = a + \varepsilon, x = b - \varepsilon$ and two curves

$$M'N': \quad y = \varphi(x) + \eta, \qquad a \le x \le b,$$
$$P'Q': \quad y = \psi(x) - \eta, \qquad a \le x \le b,$$

where ε, η are sufficiently small positive numbers. Let $M_1'N_1'Q_1'P_1'M_1'$ be the boundary of the domain which is bounded by these two straight lines and two curves, then

$$\int_{M_1'N_1'Q_1'P_1'M_1'} f(z)\,dz = 0$$

since U is a simple connected domain.

Fix ε, and let $\eta \to 0$, then

$$\int_{M_1'N_1'} f(z)\,dz \longrightarrow \int_{M_1 N_1} f(z)\,dz, \qquad \int_{Q_1'P_1'} f(z)\,dz \longrightarrow \int_{Q_1 P_1} f(z)\,dz,$$

$$\int_{P_1'M_1'} f(z)\,dz \longrightarrow \int_{P_1 M_1} f(z)\,dz, \qquad \int_{N_1'Q_1'} f(z)\,dz \longrightarrow \int_{N_1 Q_1} f(z)\,dz$$

because $f(z)$ is uniformly continuous on $\overline{U}$. Thus

$$\int_{M_1 N_1 Q_1 P_1 M_1} f(z)\,dz = 0.$$

Similarly, let $\varepsilon \to 0$, then

$$\int_{M_1 N_1} f(z)\,dz \longrightarrow \int_{MN} f(z)\,dz, \qquad \int_{Q_1 P_1} f(z)\,dz \longrightarrow \int_{QP} f(z)\,dz.$$

We have proved (2.5) if we can prove

$$\int_{P_1 N_1} f(z)\,dz \longrightarrow \int_{PM} f(z)\,dz, \qquad \int_{N_1 Q_1} f(z)\,dz \longrightarrow \int_{NQ} f(z)\,dz$$

when $\varepsilon \to 0$. We prove the last limit only, the proof of the first limit is similar.

Let

$$y_\varepsilon = \max(\varphi(b), \varphi(b-\varepsilon)), \qquad Y_\varepsilon = \max(\psi(b), \psi(b-\varepsilon)),$$

then

$$\int_{NQ} f(z)\,dz = i \int_{\varphi(b)}^{\psi(b)} f(b + i\,y)\,dy$$

$$= i\left(\int_{\varphi(b)}^{y_\varepsilon} + \int_{y_\varepsilon}^{Y_\varepsilon} + \int_{Y_\varepsilon}^{\psi(b)} \right) f(b + i\,y)\,dy,$$

$$\int_{N_1 Q_1} f(z)\,dz = i \int_{\varphi(b-\varepsilon)}^{\psi(b-\varepsilon)} f(b - \varepsilon + i\,y)\,dy$$

$$= i\left(\int_{\varphi(b-\varepsilon)}^{y_\varepsilon} + \int_{y_\varepsilon}^{Y_\varepsilon} + \int_{Y_\varepsilon}^{\psi(b-\varepsilon)} \right) f(b - \varepsilon + i\,y)\,dy.$$

We have

$$\int_{NQ} f(z)\,dz - \int_{N_1 Q_1} f(z)\,dz = i \int_{y_\varepsilon}^{Y_\varepsilon} (f(b+i\,y) - f(b-\varepsilon+i\,y))\,dy + i\,S(\varepsilon), \quad (2.6)$$

where

$$S(\varepsilon) = \left(\int_{\varphi(b)}^{y_\varepsilon} + \int_{Y_\varepsilon}^{\psi(b)} \right) f(b + iy)\,dy - \left(\int_{\varphi(b-\varepsilon)}^{y_\varepsilon} + \int_{Y_\varepsilon}^{\psi(b-\varepsilon)} \right) f(b - \varepsilon + iy)\,dy.$$

The first term in the right-hand side of (2.6) tends to zero when $\varepsilon \to 0$ because $f(z)$ is uniformly continuous on $\overline{U}$. Meanwhile $y_\varepsilon, Y_\varepsilon$ tend to $\varphi(b), \psi(b)$ respectively when $\varepsilon \to 0$. Thus the four integrals in $S(\varepsilon)$ tend to zero when $\varepsilon \to 0$. We have

$$\int_{N_1 Q_1} f(z)\,\mathrm{d}z \longrightarrow \int_{NQ} f(z)\,\mathrm{d}z \quad \text{when} \quad \varepsilon \to 0.$$

We have proved Theorem 3' when U is a special domain as we described.

For any domain U, we may add finite number of lines which are parallel with y-axis, and divide U into the sum of finite number of the special domains as we mentioned above. The integrations on these lines cancel each other. We complete the proof of Theorem 3'.

Of course Theorem 3' is true when the domain is multiple connected. It need only to add some curves, they divide the domain into the sum of finite number of simply connected domains, and the integrations on these curves cancel each other.

We may state the previous result as follows. Let $\gamma_0, \gamma_1, \cdots, \gamma_n$ be $n+1$ rectifiable curves, and let $\gamma_1, \cdots, \gamma_n$ be contained in γ_0, and each curve of $\gamma_1, \cdots, \gamma_n$ be out of others. Let U be the domain bounded by $\gamma_0, \gamma_1, \cdots, \gamma_n$, i.e., the boundary of U be $\gamma_0, \gamma_1, \cdots, \gamma_n$. If $f(z)$ is holomorphic on U, and continuous on $\overline{U}$, then

$$\int_{\partial U} f(z)\,\mathrm{d}z = 0.$$

From Theorem 3', if $f(z)$ is holomorphic on U, z_0, z are any two points in U, we may define the integral of $f(z)$ as

$$F(z) = \int_{z_0}^{z} f(\zeta)\,\mathrm{d}\zeta.$$

This integral is independent of the choice of the path. Obviously, $F'(z) = f(z)$.

§ 2.3 Taylor Series and Liouville Theorem

Starting from the Cauchy integral formula and the Cauchy integral theorem, we will get a series of important consequences. This section and the later sections in this chapter will state and prove these important consequences.

Theorem 5 Let $f(z)$ be a holomorphic function on $U \subseteq C$, and continuous on $\overline{U}$, then at any point z in U, the derivatives of all orders of $f(z)$

exist, and

$$f^{(n)}(z) = \frac{n!}{2\pi i} \int_{\partial U} \frac{f(\zeta)}{(\zeta - z)^{n+1}} \, d\zeta, \qquad n = 1, 2, \cdots .\tag{3.1}$$

If $z \in U$, $\overline{D}(z_0, r) = \{z \mid |z - z_0| \leq r\} \subset U$, then $f(z)$ can be expanded as a Taylor series

$$f(z) = \sum_{j=0}^{\infty} a_j (z - z_0)^j \tag{3.2}$$

on $D(z_0, r)$. This series converges uniformly and absolutely on $\overline{D}(z_0, r)$, and

$$a_j = \frac{1}{2\pi i} \int_{\partial U} \frac{f(\zeta) \, d\zeta}{(\zeta - z_0)^{j+1}}. \tag{3.3}$$

Proof   Let $z_0 \in U$, we may find r, so that $D(z_0, r) \subset U$. By Theorem 2', if $z \in D(z_0, r)$, then

$$f(z_0) = \frac{1}{2\pi i} \int_{\partial U} \frac{f(\zeta)}{\zeta - z_0} \, d\zeta$$

and

$$f(z) = \frac{1}{2\pi i} \int_{\partial U} \frac{f(\zeta)}{\zeta - z} \, d\zeta.$$

We have

$$f(z) - f(z_0) = \frac{1}{2\pi i} \int_{\partial U} \left(\frac{1}{\zeta - z} - \frac{1}{\zeta - z_0} \right) f(\zeta) \, dz$$

$$= \frac{z - z_0}{2\pi i} \int_{\partial U} \frac{f(\zeta) \, d\zeta}{(\zeta - z)(\zeta - z_0)},$$

and

$$\frac{f(z) - f(z_0)}{z - z_0} = \frac{1}{2\pi i} \int_{\partial U} \frac{f(\zeta) \, d\zeta}{(\zeta - z)(\zeta - z_0)}.$$

Thus

$$\frac{f(z) - f(z_0)}{z - z_0} - \frac{1}{2\pi i} \int_{\partial U} \frac{f(\zeta) \, d\zeta}{(\zeta - z)^2}$$

$$= \frac{1}{2\pi i} \int_{\partial U} \frac{f(\zeta)}{\zeta - z_0} \left(\frac{1}{\zeta - z} - \frac{1}{\zeta - z_0} \right) d\zeta$$

$$= \frac{z - z_0}{2\pi i} \int_{\partial U} \frac{f(\zeta) \, d\zeta}{(\zeta - z)(\zeta - z_0)^2}. \tag{3.4}$$

Let d denote the distance (it means the shortest distance) from z_0 to ∂U. Let $r = \frac{d}{2}$, then

$$|\zeta - z| = |(\zeta - z_0) - (z - z_0)| \geq |\zeta - z_0| - |z - z_0| \geq d - \frac{d}{2} = \frac{d}{2}.$$

Of course, $|\zeta - z_0| \geq d$. We have

$$\left| \int_{\partial U} \frac{f(\zeta)\,\mathrm{d}\zeta}{(\zeta - z)(\zeta - z_0)^2} \right| \leq \frac{ML}{\frac{d}{2} \cdot d^2} = \frac{2ML}{d^3},$$

where $M = \max\limits_{\zeta \in \partial U} |f(\zeta)|$, L=the length of ∂U. Let $z \to z_0$ in (3.4), we obtain

$$f'(z_0) = \frac{1}{2\pi i} \int_{\partial U} \frac{f(\zeta)\mathrm{d}\zeta}{(\zeta - z_0)^2}.$$

We have proved (3.1) when $n = 1$.

Assume (3.1) holds for $n = k \geq 1$,

$$f^{(k)}(z) = \frac{k!}{2\pi i} \int_{\partial U} \frac{f(\zeta)\,\mathrm{d}\zeta}{(\zeta - z)^{k+1}}.$$

We know that

$$\frac{1}{\zeta - z} = \frac{1}{\zeta - z_0} \cdot \frac{1}{1 - \dfrac{z - z_0}{\zeta - z_0}}$$

and $\left| \frac{z - z_0}{\zeta - z_0} \right| < 1$ since $|z - z_0| < r \leq |\zeta - z_0|$. Thus

$$\frac{1}{\zeta - z} = \frac{1}{\zeta - z_0} \sum_{j=0}^{\infty} \left(\frac{z - z_0}{\zeta - z_0} \right)^j. \tag{3.5}$$

We have

$$f^{(k)}(z) = \frac{k!}{2\pi i} \int_{\partial U} \frac{f(\zeta)}{(\zeta - z_0)^{k+1}} \left(1 + \frac{z - z_0}{\zeta - z_0} + \cdots \right)^{k+1} \mathrm{d}\zeta$$

$$= f^{(k)}(z_0) + \frac{(k+1)!}{2\pi i} \int_{\partial U} \frac{f(\zeta)(z - z_0)\,\mathrm{d}\zeta}{(\zeta - z_0)^{k+2}} + O(|z - z_0|^2).$$

Hence

$$\frac{f^{(k)}(z) - f^k(z_0)}{z - z_0} = \frac{(k+1)!}{2\pi i} \int_{\partial U} \frac{f(\zeta)\,\mathrm{d}\zeta}{(\zeta - z_0)^{k+2}} + O(|z - z_0|).$$

Let $z \to z_0$, we obtain (3.1) for $n = k + 1$,

$$f^{(k+1)}(z_0) = \frac{(k+1)!}{2\pi i} \int_{\partial U} \frac{f(\zeta)\,d\zeta}{(\zeta - z_0)^{k+2}}.$$

By mathematical induction, (3.1) holds for any $n = 1, 2, \cdots$.

Substituting (3.5) into $f(z) = \frac{1}{2\pi i} \int_{\partial U} \frac{f(\zeta)\,d\zeta}{\zeta - z}$, we have

$$f(z) = \frac{1}{2\pi i} \int_{\partial U} \frac{f(\zeta)}{\zeta - z_0} \sum_{j=0}^{\infty} \left(\frac{z - z_0}{\zeta - z_0}\right)^j d\zeta.$$

By Theorem 4 of Chapter I, the integral sign and $\sum\limits_{j=0}^{\infty}$ can exchange in previous equality. Hence,

$$f(z) = \sum_{j=0}^{\infty} (z - z_0)^j \frac{1}{2\pi i} \int_{\partial U} \frac{f(\zeta)\,d\zeta}{(\zeta - z_0)^{j+1}}$$

$$= \sum_{j=0}^{\infty} \frac{f^{(j)}(z_0)}{j!} (z - z_0)^j$$

by (3.1). We have proved (3.2) and (3.3).

From Theorem 5, we know that, in the case of function of complex variable, the derivatives of all orders exist and the function can be expanded as a Taylor series if the derivative of first order exist. This property does not hold in the case of function of real variable. This is one essential difference between function of complex variable and function of real variable. In §1.3 of Chapter I, we defined that a function $f(z)$ is holomorphic on $U \subseteq C$ if the derivative of $f(z)$ at any point in U exists. From Theorem 5, we may define a function $f(z)$ holomorphic on $U \subseteq C$ if $f(z)$ can be expanded as a convergence power series at a neighborhood of each point in U. These two definitions are equivalent to each other.

From Theorem 5, we immediately obtain the following result.

Theorem 6   (1) Cauchy inequality. Let $f(z)$ be holomorphic on $U \subseteq \mathbb{C}$, and let $z_0 \in U$, $\overline{D}(z_0, R) \subseteq U$, then

$$|f^{(j)}(z_0)| \leq \frac{j!M}{R^j}, \qquad j = 1, 2, \cdots \tag{3.6}$$

holds where $M = \max\limits_{z \in \overline{D}(z_0, R)} |f(z)|$.

(2) Let $U \subseteq \mathbb{C}$ be a domain, and let K be a compact set in U, and let V be a neighborhood of K which is relatively compact in U (i.e. $\overline{V}$ is a compact subset in U), then for any holomorphic function $f(z)$ on U, there exists constants c_n $(n = 1, 2, \cdots)$ such that

$$\sup_{z \in K} |f^{(n)}(z)| \le c_n \|f\|_{L(v)}, \qquad n = 1, 2, \cdots, \tag{3.7}$$

where $\|f\|_{L(v)}$ is the L-norm of f on V,

$$\frac{1}{A(V)} \iint\limits_V |f(\zeta)| \, dA,$$

where $A(V)$ is the area of V.

The Cauchy inequality gives the estimate of the modular of derivatives of all orders of the holomorphic function at one point, and formula (3.7) at Theorem 6(2) gives the estimate of the modular of derivatives of all order of the holomorphic function at a compact set.

Proof   Theorem 6(1) is obvious, we need only to prove Theorem 6(2).

We construct a C^∞ function ψ on V so that ψ has a compact support on V, and $\psi = 1$ at a neighborhood of K which is contained in V. The function ψ exists (cf. Appendix of this Chapter). Using Theorem 1 (Pompeiu formula) to ψf, then

$$\psi(z)f(z) = \frac{1}{2\pi i} \int_{\partial U} \frac{\psi(\zeta)f(\zeta)\,d\zeta}{\zeta - z} + \frac{1}{2\pi i} \iint\limits_U \frac{\partial(\psi f)}{\partial \overline{\zeta}} \frac{d\zeta \wedge d\overline{\zeta}}{\zeta - z}.$$

Since f is holomorphic on U, we have $\frac{\partial(\psi f)}{\partial \overline{\zeta}} = f\frac{\partial \psi}{\partial \overline{\zeta}}$. From the fact that the support of $\psi(\zeta)$ is contained in V, and V is relative compact on U, we have

$$\psi(z)f(z) = \frac{1}{2\pi i} \iint\limits_U f\frac{\partial \psi}{\partial \overline{\zeta}} \frac{d\zeta \wedge d\overline{\zeta}}{\zeta - z}.$$

If the support of $\frac{\partial \psi}{\partial \overline{\zeta}}$ is K_1, then K_1 is a compact subset in V, and the distance between K and K_1, $d(K, K_1) > 0$.

If $z \in K$, then

$$f(z) = \frac{1}{2\pi i} \iint\limits_{K_1} f(\zeta) \frac{\partial \psi(\zeta)}{\partial \overline{\zeta}} \frac{d\zeta \wedge d\overline{\zeta}}{\zeta - z}.$$

Differentiating both sides n times with respect to z, we have

$$f^{(n)}(z) = \frac{n!}{2\pi i} \iint\limits_{K_1} f(\zeta) \frac{\partial \psi(\zeta)}{\partial \overline{\zeta}} \frac{d\zeta \wedge d\overline{\zeta}}{(\zeta - z)^{n+1}}.$$

Hence

$$|f^{(n)}(z)| \leq \frac{n!}{2\pi} \iint\limits_{K_1} |f(\zeta)| \left| \frac{\partial \psi(\zeta)}{\partial \overline{\zeta}} \right| \frac{d\zeta \wedge d\overline{\zeta}}{|\zeta - z|^{n+1}}.$$

Since $d(K, K_1) > 0$, there exists a constant c_1 so that $\frac{1}{|\zeta - z|} < c_1$ holds for any $z \in K$ and $\zeta \in K_1$. Obviously, $\left| \frac{\partial \psi(\zeta)}{\partial \overline{\zeta}} \right|$ is bounded on K_1. Thus, there exists a constant c_n' such that

$$|f^{(n)}(z)| \leq c_n' \iint\limits_{K_1} |f(\zeta)| \, |d\zeta \wedge d\overline{\zeta}|$$

$$\leq c_n' \iint\limits_{V} |f(\zeta)| \, |d\zeta \wedge d\overline{\zeta}| = c_n \|f\|_{L(v)},$$

where c_n', c_n are constants dapending on n only.

We have proved (3.7).

From Theorem 6(2), we have the following result.

Corollary 1 The assumptions are the same as Theorem 6(2), then for any holomorphic function on U, there exists constants c_n $(n = 1, 2, \cdots)$ such that

$$\sup_{z \in K} |f^{(n)}(z)| \leq c_n \sup_{z \in V} |f(z)|, \qquad n = 1, 2, \cdots$$

holds.

We may prove Corollary 1 from Theorem 5 directly.

From Theorem 5, we obtain the converse thorem of Theorem 3' (the Cauchy-Goursat integration theorem).

Theorem 7 (Morera theorem) Let $f(z)$ be a continuous function on U. If the integration of $f(z)$ along any closed rectifiable curve inside U is zero, then $f(z)$ is holomorphic on U.

Proof Since the integration of $f(z)$ along any closed rectifiable curve inside U is zero, we may define

$$F(z) = \int_{z_0}^{z} f(\zeta) \, \mathrm{d}\zeta, \qquad z \in U$$

along any path inside U and connecting z_0 and z. Clearly $F'(z) = f(z)$. Therefore $F(z)$ is a holomorphic function on U, by Theorem 5 the second derivative of $F(z)$, that is, the derivative of $f(z)$, $f'(z)$ exist. Hence $f(z)$ is a holomorphic function on U.

Moreover, from Theorem 6, we have the following important theorem.

Theorem 8 (Liouville theorem) If $f(z)$ is a bounded holomorphic function on the whole plane $\mathbb{C}$, then f is a constant.

Proof Assume $|f(z)| \leq M$ when $z \in \mathbb{C}$. Fix $z_0 \in \mathbb{C}$, denote $D(z_0, R)$ as a disk centered at z_0 with radius R. Then $|f'(z_0)| \leq \frac{M}{R}$ by (3.6) with $j = 1$. Letting $R \to 0$, we have $f'(z_0) = 0$. Since z_0 is an arbitrary point in $\mathbb{C}$, we obtain $f'(z) = 0$ for any $z \in \mathbb{C}$. Thus $f(z)$ is a constant on $\mathbb{C}$.

Liouville theorem tells us: The function which is holomorphic and bounded on the whole complex plane $\mathbb{C}$ is a constant function only. We will discuss this theorem again in Chapter V.

Finally, we prove the following theorem.

Theorem 9 (Riemann theorem) Let $\tilde{D}(z_0, r) = D(z_0, r) \backslash \{z_0\}$, and let F be holomorphic and bounded on $\tilde{D}(z_0, r)$, then F can be analytically continued to $D(z_0, r)$. That is, we may define a holomorphic function f on $D(z_0, r)$ with $f|_{\tilde{D}(z_0, r)} = F$.

Proof Without loss of generality, we may assume $z_0 = 0$. Defining

$$G(z) = \begin{cases} z^2 F(z), & \text{when } z \in \tilde{D}(0, r), \\ 0, & \text{when } z = 0, \end{cases}$$

we have

$$\lim_{z \to 0} \frac{G(z) - 0}{z} = \lim_{z \to 0} \frac{z^2 F(z) - 0}{z} = \lim_{z \to 0} z F(z) = 0.$$

It is $G'(0) = 0$. When $z \neq 0$, we have $G'(z) = z^2 F'(z) + 2z F(z)$, hence $G'(z) \to 0$ when $z \to 0$. Thus $G(z)$ is continuous and differentiable on $D(0, r)$, and satisfies the Cauchy -Riemann equation. By Theorem 1 in Chapter I §1.3, $G(z)$ is holomorphic on $D(0, r)$. Hence we may expand $G(z)$ at $z = 0$ as a Taylor series

$$G(z) = 0 + 0 \cdot z + a_2 z^2 + a_3 z^3 + \cdots, \tag{3.8}$$

and the series converges on $D(0, r)$. Define

$$f(z) = \frac{G(z)}{z^2} = a_2 + a_3 z + \cdots, \tag{3.9}$$

the series (3.9) and (3.8) have the same radius of convergence by (6.2) of Theorem 3 in Chapter I. Thus $f(z)$ is holomorphic on $D(0, r)$, and $f(z) = F(z)$ when $z \in \widetilde{D}(0, r)$.

§ 2.4   Some Results on Zero Points

Let $f(z)$ be a holomorphic function on $U \subseteq \mathbb{C}$. z_0 is called a **zero point** of $f(z)$ if $z_0 \in U$ and $f(z_0) = 0$. $f(z)$ has a **zero point of order m** at $z = z_0$ if $f(z)$ has an expansion

$$a_m (z - z_0)^m + a_{m+1} (z - z_0)^{m+1} + \cdots, \qquad a_m \neq 0$$

at $z = z_0$. We may derive a series of results about the zero points from the Cauchy integral formula and the Cauchy integral theorem.

Theorem 10 (Fundamental theorem of algebra)   Let $p(z) = a_0 + a_1 z + \cdots + a_n z^n$ be a non-constant polynomial of degree n, then there exists at least one point z_0 such that $p(z_0) = 0$.

The point z_0 is the root of equation $p(z) = 0$.

Proof   If it is not true, then $f(z) = \frac{1}{p(z)}$ is holomorphic on $\mathbb{C}$. The function $f(z)$ is bounded on $\mathbb{C}$ since $p(z) \to \infty$ when $z \to \infty$. By Theorem 8 (Liouville theorem), $f(z)$ is a constant function, and $p(z)$ is a constant polynomial. It is impossible.

Theorem 11   Let $f(z)$ be holomorphic on $U \subseteq \mathbb{C}$, then the set of zero points of $f : \{z \in U \mid f(z) = 0\}$, has no limiting point on U, unless $f(z)$ is identically equal to zero on U.

Proof   If it is not true, let $z_1, z_2, \cdots, z_n, \cdots$ be the zero points of $f(z)$ on U, and let $z_0 \in U$ be the limiting point. Without loss of generality, we may assume $z_0 = 0$. We may expand $f(z)$ at $z = 0$ as

$$f(z) = a_0 + a_1 z + a_2 z^2 + \cdots$$

since $0 \in U$ and $f(z)$ is holomorphic on U. Recall the sequence $\{z_n\}$ $(n =$

$1, 2, \cdots$) is the zero points of $f(z)$, we have $f(z_n) = 0$. Thus

$$\lim_{n \to \infty} f(z_n) = f(\lim_{n \to \infty} z_n) = f(0) = 0.$$

We have $a_0 = 0$, and

$$f(z) = a_1 z + a_2 z^2 + \cdots.$$

It yields $a_1 = \frac{f(z)}{z} + O(z)$. Let $z = z_n$, then $a_1 = \frac{f(z_n)}{z_n} + O(z_n) = O(z_n)$, and $a_1 = 0$ when $n \to \infty$. We may prove $a_2 = a_3 = \cdots = a_n = \cdots = 0$ by the same process, then all coefficients of Taylor series are zero. Hence $f(z) = 0$. We conclude that the set $\{z \in U \mid f(z) = 0\}$ has no limiting point on U if $f(z)$ is not identically equal to zero on U.

From Theorem 11, we have the following consequence. Suppose $h_1(z)$ and $h_2(z)$ are two holomorphic functions on domain $U \subseteq C$, E is a set on U, E has limiting points and all these limiting points are inner points of U. If $h_1(z) = h_2(z)$ when $z \in E$, then $h_1(z) = h_2(z)$ when $z \in U$. It means that the values of a holomorphic function on U can be decided by the values of the function on a set for which each point of the set and the limiting points of the set are belong to U. For example, $\sin^2 z + \cos^2 z = 1$ holds for any complex number because it holds for any real number. Similarly, some trigonometric identities hold for any complex number if they hold for any real number.

Theorem 12 (Principle of argument) Let $f(z)$ be a holomorphic function on a domain $U \subseteq \mathbb{C}$, and let $\gamma \subset U$ be a simple closed curve with positive orientation, and γ may shrink to a point in U continuously. If $f(z)$ is non-zero when $z \in \gamma$, then $f(z)$ has a finite number of zero points inside the domain which is bounded by γ, and the number of zero points (counting multiplicity) equals

$$k = \frac{1}{2\pi i} \int_\gamma \frac{f'(z)}{f(z)} \, dz. \tag{4.1}$$

If we let $w = f(z)$, (4.1) is

$$k = \frac{1}{2\pi i} \int_\Gamma \frac{dw}{w} = \frac{1}{2\pi} \Delta_\Gamma \operatorname{Arg} w,$$

where Γ is the image of γ under $w = f(z)$ and $\Delta_\Gamma \operatorname{Arg} w$ denote the change of argument of w on Γ.

It means that when z moves around γ with positive orientation once the number of cycles on Γ in which $w = f(z)$ rotates with respect to origin with positive orientation equals the number of zero points of f inside the domain which is bounded by γ. This is the reason why we call Theorem 12 as the principle of argument.

Proof of Theorem 12   Assume $f(z)$ have zero points $z_1, z_2, \cdots, z_n$ with multiplicity $k_1, \cdots, k_n$ respectively in a domain which is bounded by γ. At each zero point z_i $(i = 1, \cdots, n)$, we construct a circle γ_i with radius $\varepsilon_i > 0$ and centred at z_i, such that all γ_i $(i = 1, \cdots, n)$ are inside γ and mutually non-intersect to each other. Then

$$\frac{1}{2\pi i} \int_\gamma \frac{f'(z)}{f(z)}\, dz = \sum_{i=1}^n \frac{1}{2\pi i} \int_{\gamma_i} \frac{f'(z)}{f(z)}\, dz.$$

Since $f(z)$ has a zero point at $z = z_i$ with multiplicity k_i, $f(z)$ can be expressed as

$$f(z) = (z - z_i)^{k_i} h_i(z),$$

where $h_i(z) \neq 0$ for any inner point z of the domain which is bounded by γ_i. Hence

$$f'(z) = k_i(z - z_i)^{k_i - 1} h_i(z) + (z - z_i)^{k_i} h_i'(z).$$

We have

$$\frac{f'(z)}{f(z)} = \frac{k_i}{z - z_i} + \frac{h_i'(z)}{h_i(z)}.$$

It follows

$$\frac{1}{2\pi i} \int_{\gamma_i} \frac{f'(z)}{f(z)}\, dz = k_i.$$

That is

$$\frac{1}{2\pi i} \int_\gamma \frac{f'(z)}{f(z)}\, dz = \sum k_i = k.$$

Theorem 13 (Hurwitz theorem)   Let f_j $(j = 1, 2, \cdots)$ be a sequence of holomorphic functions on $U \subseteq \mathbb{C}$, and let it converge to a function f uniformly on any compact subset in U. Moreover, if every f_j $(j = 1, 2, \cdots)$ has no zero point on U, then f either has no zero point on U or identically equals to zero.

Proof   For any point $z \in U$, we may take a simple closed curve γ in U, such that z belongs to the domain which is bounded by γ, we have

$$f_j(z) = \frac{1}{2\pi i} \int_\gamma \frac{f_j(\zeta)\, d\zeta}{\zeta - z}.$$

since f_j is holomorphic on U. From the uniform convergence of f_j $(j = 1, 2, \cdots)$ on any compact subset in U, we have

$$\lim_{j \to \infty} f_j(z) = \lim_{j \to \infty} \frac{1}{2\pi i} \int_\gamma \frac{f_j(\zeta) \, d\zeta}{\zeta - z} = \frac{1}{2\pi i} \int_\gamma \lim_{j \to \infty} f_j(\zeta) \frac{d\zeta}{\zeta - z}$$

(We may change the order of limit and integration due to the uniform convergence on any compact subset in U. It is a consequence of the Theorem 4 of Chapter I also). That is,

$$f(z) = \frac{1}{2\pi i} \int_\gamma \frac{f(\zeta) \, d\zeta}{\zeta - z}.$$

Thus $f(z)$ is a holomorphic function. Similarly, we may prove that $\{f'_j(z)\}$ converges uniformly to $f'(z)$ on any compact subset in U.

If $f(z) \not\equiv 0$, then all the zero points of $f(z)$ are discrete by Theorem 11. Let γ be a curve which does not pass any zero point of f, then

$$\frac{1}{2\pi i} \int_\gamma \frac{f'_j(\zeta)}{f_j(\zeta)} \, d\zeta \longrightarrow \frac{1}{2\pi i} \int_\gamma \frac{f'(\zeta)}{f(\zeta)} \, d\zeta$$

when $j \to \infty$. By the hypothesis and Theorem 12, we know that

$$\frac{1}{2\pi i} \int_\gamma \frac{f'_j(z)}{f_j(z)} \, d\zeta = 0.$$

Hence

$$\frac{1}{2\pi i} \int_\gamma \frac{f'(z)}{f(z)} \, d\zeta = 0.$$

Therefore, $f(z)$ has no zero point on U.

Theorem 14 (Rouché theorem) Let $f(z), g(z)$ be holomorphic functions on $U \subseteq \mathbb{C}$, and let γ be a rectifiable simple closed curve in U. Moreover, if f and g satisfy the inequality

$$|f(z) - g(z)| < |f(z)| \tag{4.2}$$

for any $z \in \gamma$, then f and g have the same number of zero points on the domain which is bounded by γ.

Proof We know that $|f(z)| > 0$ and $g(z) \neq 0$ when $z \in \gamma$. If it is not true, then there exists a point $z_0 \in \gamma$, such that $g(z_0) = 0$. It implies a contradiction, $|f(z_0)| > |f(z_0)|$.

Let N_1 and N_2 be the number of zero points of f and g respectivrely on the domain which is bounded by γ, then

$$N_1 = \frac{1}{2\pi i} \int_\gamma \frac{f'(z)}{f(z)}\, dz, \qquad N_2 = \frac{1}{2\pi i} \int_\gamma \frac{g'(z)}{g(z)}\, dz$$

by Theorem 12 (Principle of argument). We have

$$N_2 - N_1 = \frac{1}{2\pi i} \int_\gamma \left(\frac{g'(z)}{g(z)} - \frac{f'(z)}{f(z)}\right) dz$$

$$= \frac{1}{2\pi i} \int_\gamma \frac{fg' - gf'}{gf}\, dz$$

$$= \frac{1}{2\pi i} \int_\gamma \frac{(g/f)'}{g/f}\, dz.$$

Denote $F(z) = g(z)/f(z)$, then

$$N_2 - N_1 = \frac{1}{2\pi i} \int_\gamma \frac{F'(z)}{F(z)}\, dz.$$

The inequality (4.2) can be expressed as $|F(z) - 1| < 1$. $w = F(z)$ maps γ to Γ, and Γ neither pass origin nor contains origin since Γ is inside the domain $|w - 1| < 1$. We have $\int_\Gamma \frac{dw}{w} = 0$ by Theorem 3 (Cauchy integration Theorem).

That is, $N_1 = N_2$.

If $p(z) = a_n z^n + a_{n-1} z^{n-1} + \cdots + a_0$ is a polynomial of degree n, $p(z)$ has a zero point z_0 at least, $p(z_0) = 0$. Now, we can prove that $p(z)$ has exact n zero points if $a_n \neq 0$ by Rouché theorem.

Let $g(z) = a_n z^n$, then

$$|p(z) - g(z)| = |a_{n-1} z^{n-1} + \cdots + a_n| < |g(z)| = |a_n||z|^n = |a_n|R^n$$

on $|z| = R$ when R is sufficiently large. Thus $p(z)$ and $g(z)$ have the same number of zero points on $|z| < R$ by Rouché theorem. Obviously $a_n z^n$ has n zero points, and hence $p(z)$ has exact n zero points.

As a consequence of Roché theorem, we have

Theorem 15   Let $f(z)$ be holomorphic on $U \subseteq \mathbb{C}$, and let $w_0 = f(z_0), z_0 \in U$. If z_0 is a zero point of function $f(z) - w_0$ with multiplicity m, then for sufficiently small $\rho > 0$, there exists a $\delta > 0$ such that for every point $A \in D(w_0, \delta), f(z) - A$ has m zero points on $D(z_0, \rho)$.

Proof If z_0 is a zero point of multiplicity m of function $f(z) - f(z_0)$. By Theorem 11, there exists a ρ_0, such that $f(z) - f(z_0)$ has no zero point on $\overline{D}(z_0, \rho) \subset U$ except $z = z_0$, and $|f(z) - f(z_0)| \geq \delta(\delta > 0)$ holds on $|z - z_0| = \rho$. Then for any point $A \in D(w_0, \delta), |A - w_0| < |f(z) - f(z_0)|$ holds on $|z - z_0| = \rho$. That is, $|(f(z) - f(z_0)) - (f(z) - A)| < |f(z) - f(z_0)|$ holds on $|z - z_0| = \rho$. By Rouché theorem, $f(z) - A$ and $f(z) - f(z_0)$ have the same number of zero points on $D(z_0, \rho)$. Of course, $f(z) - f(z_0)$ has zero point at $z = z_0$ with multiplicity m, and no other zero point on $D(z_0, \rho)$. Hence $f(z) - A$ has m zero points on $D(z_0, \rho)$.

We will discuss the theory of zero points again in the later.

§ 2.5   Maximum Modulus Principle, Schwarz Lemma, Group of Holomorphic Automorphism

Another important consequence of Cauchy integral formula is the maximum modulus principle. This is a very useful theorem.

Before we state this theorem, we prove the mean-value property of holomorphic functions.

Let $f(z)$ be a holomorphic function on $U \subseteq \mathbb{C}$, and $z_0 \in U$. If $r > 0$, and $\overline{D}(z_0, r) \subset U$, then

$$f(z_0) = \frac{1}{2\pi i} \int_{\partial D(z_0, r)} \frac{f(\zeta) \, d\zeta}{\zeta - z_0}$$

by Cauchy integral formula. The point ζ on $\partial D(z_0, r)$ can be expressed as $\zeta = z_0 + re^{it}$ $(0 \leq t \leq 2\pi)$. Then the Cauchy integral formula can be written as

$$f(z_0) = \frac{1}{2\pi i} \int_0^{2\pi} \frac{f(z_0 + re^{it}) i re^{it} \, dt}{re^{it}}$$

$$= \frac{1}{2\pi} \int_0^{2\pi} f(z_0 + re^{it}) \, dt. \tag{5.1}$$

This is the mean value property of holomorphic functions. It means that the value of $f(z)$ at $z = z_0$ is equal to the mean value of $f(z)$ on $\partial D(z_0, r)$.

Taking real part and imaginary part on both sides of (5.1), we know that the harmonic function has the mean value property also. Conversely, we may prove that if a continuous function has the mean value property, then it is a harmonic function. We will prove it in §2.6. Thus a continuous function is a harmonic function if and only if it has the mean value property, thus we may define a harmonic function as a function having the mean value property.

Using the mean value property of holomorphic functions, we prove the following theorem.

Theorem 16 (Maximum modulus principle)   Let $f(z)$ be a holomorphic function on domain $U \subseteq \mathbb{C}$, and let $z_0 \in U$ such that $|f(z_0)| \geq |f(z)|$ holds for all $z \in U$, then $f(z)$ is a constant function.

Proof   We may assume $M = f(z_0) \geq 0$, otherwise we multiply a constant with modulus one on f. Let $S = \{z \in U \mid f(z) = f(z_0)\}$, then $S \neq \phi$ (ϕ means empty set) since $z_0 \in S$. S is closed since f is continuous. Now we prove that S is open. If $w \in S$, then there exists a $r > 0$, such that $D(w, r) \subseteq U$. Let $0 < r' < r$, then

$$M = f(w) = \left| \frac{1}{2\pi} \int_0^{2\pi} f(w + r'e^{\mathrm{i}t}) \, \mathrm{d}t \right|$$

$$\leq \frac{1}{2\pi} \int_0^{2\pi} |f(w + r'e^{\mathrm{i}t})| \, \mathrm{d}t \leq M.$$

The equality holds in the previous inequality because the left side and right side are equal, that is

$$f(w + r'e^{\mathrm{i}t}) = |f(w + r'e^{\mathrm{i}t})| = M$$

holds for all t and $0 < r' < r$. Thus

$$\{w + r'e^{\mathrm{i}t} \mid 0 \leq t \leq 2\pi, \ 0 < r' < r\} \subseteq S.$$

It means that for any point $w \in S$, there exists a small open disk such that every point of this disk belongs to S. It implies that S is open. Hence S is a non-empty, closed and open set in U. Moreover, U is connected, we have $S = U$. It yields that $f(z)$ is a constant function on U.

In the proof, we used a very powerful theorem: Any non-empty, closed and open subset of a connected set is the set itself. It remains as an exercise.

As a direct consequence of the maximum modulus principle, we have the following conclusion. Let $f(z)$ be a holomorphic function on a bounded domain $U \subseteq \mathbb{C}$, and let $f(z)$ be continuous on $\overline{U}$, then $|f(z)|$ attends its maximum value on ∂U provided $f(z)$ is not a constant function on U.

In the proof of maximum modulus principle for holomorphic function, we used the mean value property of function only. Thus the maximum modulus principle holds also for harmonic functions.

From the maximum modulus principle, we have the following very important theorem.

Theorem 17 (Schwarz lemma) Let $f(z)$ be a holomorphic function of the unit disk $D = D(0,1)$ into D, and $f(0) = 0$, then

$$|f(z)| \le |z| \quad \text{and} \quad |f'(0)| \le 1. \tag{5.2}$$

$|f(z)| = |z|$ at a point $z \in D$ or $|f'(0)| = 1$ if and only if $f(z) = e^{i\tau}z$ where $\tau \in \mathbb{R}$.

Proof Let

$$G(z) = \begin{cases} \frac{f(z)}{z}, & \text{when } z \ne 0, \\ f'(0), & \text{when } z = 0, \end{cases}$$

then $G(z)$ is holomorphic on D. Using the maximum modulus principle to the function $G(z)$ on the domain $\{z \mid |z| < 1 - \varepsilon\}(\varepsilon > 0)$, we have

$$|G(z)| \le \frac{\max\limits_{|z|=1-\varepsilon} |f(z)|}{1 - \varepsilon} < \frac{1}{1 - \varepsilon}.$$

Let $\varepsilon \to 0+$, then $|G(z)| \le 1$ when $z \in D$. Thus, $|f(z)| \le |z|$ when $z \ne 0$, and $|G(0)| = |f'(0)| \le 1$.

If $|f(z)| = |z|$ at a point $z \ne 0$, $z \in D$, then $|G(z)| = 1$ at a point $z \ne 0$, $z \in D$, so $|G(z)| = 1$ for all $z \in D$ by the maximum modulus principle. Hence $G(z) = e^{it}$, $\tau \in \mathbb{R}$, $f(z) = e^{i\tau}z$. Similarly, we may prove $f(z) = e^{i\tau}z$ when $|f'(0)| = 1$.

Using Schwarz lemma, we may obtain the group of holomorphic automorphisms of unit disk D.

Let domain $U \subseteq \mathbb{C}$, we define the group of holomorphic automorphisms of U as follows.

If a holomorphic function $f(z)$ is defined on U, and $f(z)$ maps U onto U itself univalently, then $f(z)$ is a holomorphic automorphism of U. All the holomorphic automorphisms of U form a group, it is the group of holomorphic automorphisms of U, and is denoted by $\text{Aut}\,(U)$.

Now we try to decide $\text{Aut}\,(D)$.

At first, we prove that: if $a \in D$, then $\varphi_a(\zeta) = \frac{-\zeta+a}{1-\bar{a}\zeta} \in \text{Aut}\,(D)$.

Obviously, φ_a is holomorphic on $\overline{D}, \varphi_a(a) = 0$, and $\varphi_a : \partial D \to \partial D$ due to

$$|\varphi_a(\zeta)| = \left|\frac{-\zeta + a}{1 - \overline{a}\zeta}\right| = \left|\frac{1}{\zeta} \cdot \frac{-\zeta + a}{1 - \overline{a}\zeta}\right| = \left|\frac{\zeta - a}{\overline{\zeta} - a}\right| = 1$$

when $|\zeta| = 1$. It follows that $\varphi_a(\zeta)$ maps D onto D.

Secondly, we prove that $\varphi_a(\zeta)$ is univalent on D.

If it is not true, there are $\zeta_1, \zeta_2 \in D$ such that

$$\frac{-\zeta_1 + a}{1 - \overline{a}\zeta_1} = \frac{-\zeta_2 + a}{1 - \overline{a}\zeta_2},$$

then

$$(\zeta_1 - a)(1 - \overline{a}\zeta_2) = (\zeta_2 - a)(1 - \overline{a}\zeta_1),$$

it is $(\zeta_1 - \zeta_2)(1 - |a|^2) = 0$. We have $\zeta_1 = \zeta_2$ since $|a| < 1$. We have proved $\varphi_a(\zeta) \in \text{Aut}(D)$.

Let $\xi = \varphi_a(\zeta) = \frac{-\zeta + a}{1 - \overline{a}\zeta}$, then $\xi - \overline{a}\zeta\xi = -\zeta + a$, and

$$\zeta = \frac{-\xi + a}{1 - \overline{a}\,\xi} = \varphi_a(\xi).$$

We have $(\varphi_a)^{-1} = \varphi_a$. We call φ_a a Möbius transformation. All the Möbius transformations form a group, the Möbius transformation group, which is a subgroup of $\text{Aut}(D)$. Moreover, the rotation $\xi = \rho_\tau(\zeta) = e^{i\tau}\zeta$, $\tau \in \mathbb{R}$ belongs to $\text{Aut}(D)$ also. All the rotations form a group, the rotation group, which is a subgroup of $\text{Aut}(D)$ also.

Theorem 18 (Group of holomorphic automorphisms of unit disk)
If $f \in \text{Aut}(D)$, then there exists $a \in \mathbb{C}, |a| < 1$, and $\tau \in \mathbb{R}$, such that

$$f(\zeta) = \varphi_a \circ \rho_\tau(\zeta). \tag{5.3}$$

Hence, all the elements of $\text{Aut}(D)$ are composed by Möbius transformation and rotation.

Proof   Set $f(0) = b$, and $G = \varphi_b \circ f$, then G is holomorphic and univalent on D, G maps D onto D, and

$$G(0) = \varphi_b \circ f(0) = \varphi_b(b) = 0.$$

By Theorem 17 (Schwarz lemma), $|G'(0)| \le 1$. Since G is holomorphic and univalent on D, G^{-1} exists on D. G^{-1} is holomorphic and univalent on D, and

$G^{-1}(0) = 0$. We may use Schwarz lemma to G^{-1} again to obtain

$$\left|\frac{1}{G'(0)}\right| = |(G^{-1})'(0)| \leq 1.$$

Hence $|G'(0)| = 1$. By Theorem 17, it implies $G(\zeta) = e^{i\tau}\zeta = \rho_\tau(\zeta)$, it is $\varphi_b \circ f = \rho_\tau$. Thus $f = \varphi_{-b} \circ \rho_\tau$. We have proved (5.3) if we let $-b = a$.

From Theorem 18, we have the following theorem.

Theorem 19 (Schwarz-Pick lemma) Let f be a holomorphic function on D and let f map D into D, and map $z_1, z_2 \in D$ to $w_1 = f(z_1)$, $w_2 = f(z_2)$ respectively, then

$$\left|\frac{w_1 - w_2}{1 - w_1\overline{w}_2}\right| \leq \left|\frac{z_1 - z_2}{1 - z_1\overline{z}_2}\right| \tag{5.4}$$

and

$$\frac{|dw|}{1 - |w|^2} \leq \frac{|dz|}{1 - |z|^2}, \tag{5.5}$$

equality holds if and only if $f \in \text{Aut}(D)$.

Proof Let

$$\varphi(z) = \frac{z + z_1}{1 + \overline{z}_1 z}, \qquad \psi(z) = \frac{z - w_1}{1 - \overline{w}_1 z},$$

then $\varphi, \psi \in \text{Aut}\, D$, and

$$\psi \circ f \circ \varphi(0) = \psi \circ f(z_1) = \psi(w_1) = 0.$$

Thus $\psi \circ f \circ \varphi$ satisfies all conditions in Theorem 17 (Schwarz lemma). We have

$$|(\psi \circ f \circ \varphi)(z)| \leq |z|$$

when $z \in D$ and $z \neq 0$. Let $z = \varphi^{-1}(z_2)$, then

$$|\psi \circ f(z_2)| \leq |\varphi^{-1}(z_2)|.$$

It is (5.4).

When $z = 0$, we have

$$|(\psi \circ f \circ \varphi)'(0)| \leq 1$$

by Theorem 17 (Schwarz lemma), it is

$$|\psi'(w_1)f'(z_1)\varphi'(0)| \le 1.$$

We know that

$$\varphi'(z) = \frac{1 - z_1\bar{z}_1}{(1 + \bar{z}_1 z)^2}, \qquad \varphi'(0) = 1 - |z_1|^2;$$

$$\psi'(z) = \frac{1 - w_1\bar{w}_1}{(1 - \bar{w}_1 w)^2}, \qquad \psi'(w_1) = \frac{1}{1 - |w_1|^2}.$$

Hence $|f'(z_1)| \le \frac{1-|w_1|^2}{1-|z_1|^2}$. It is (5.5).

By Theorem 17, equality holds if and only if

$$\psi \circ f \circ \varphi(z) = e^{\mathrm{i}\tau}z = \rho_\tau(z).$$

It is,

$$f = \psi^{-1} \circ \rho_\tau \circ \varphi^{-1} \in \mathrm{Aut}\,(D).$$

The theorem have proved.

In fact, we may define a metric (hyperbolic metric, Poincaré metric, cf. Chapter V, §5.1) on D, by

$$\mathrm{d}_z s^2 = \frac{|\mathrm{d}z|^2}{(1 - |z|^2)^2},$$

then (5.5) becomes $\mathrm{d}_w s^2 \le \mathrm{d}_z s^2$. Hence, we may restate Theorem 19 as follows. If $w = f(z)$ is holomorphic function of D into D, then the Poincaré metric is non-increasing. The Poincaré metric is invariant if and only if $f \in \mathrm{Aut}\,(D)$. Theorem 19 gives the clear geometrical meaning of Schwarz lemma.

§ 2.6   Integral Representation of Holomorphic Function

Cauchy integral formula (2.1) is the one of the integral representation of holomorphic functions. Denote $\frac{1}{2\pi\mathrm{i}}\frac{1}{\zeta-z} = H(\zeta, z)$, and call $H(\zeta, z)$ as **Cauchy kernel**. (2.1) can be written as

$$f(z) = \int_{\partial U} f(\zeta)H(\zeta, z)\,\mathrm{d}\zeta.$$

It means that, the value of $f(z)$ at a point z can be expressed as the integration of the Cauchy kernel $H(\zeta, z)$ with the values of f on the boundary ∂U of U.

Some other integral representations can be derived. First, we give the integral representation of harmonic functions.

Let $U(z)$ be a harmonic function on unit disk D, and continuous on $\overline{D}$, then

$$\frac{1}{2\pi} \int_0^{2\pi} U(e^{i\psi}) \, d\psi = U(0) \tag{6.1}$$

by the mean value property of harmonic functions (cf. §2.5).

If $a \in D$, then $w = \frac{z-a}{1-\bar{a}z} \in \mathrm{Aut}\,(D)$. We already know that it maps ∂D onto ∂D (cf. §2.5). Let $U(w) = u(z)$, then $u(z)$ is a harmonic function on D, and $U(0) = u(a)$. If $e^{i\psi} = \frac{e^{i\tau}-a}{1-\bar{a}e^{i\tau}}$, then

$$d\psi = \frac{1-|a|^2}{|1-\bar{a}e^{i\tau}|^2} \, d\tau.$$

Substituting it into (6.1), we have

$$\frac{1}{2\pi} \int_0^{2\pi} u(e^{i\tau}) \frac{1-|a|^2}{|1-\bar{a}e^{i\tau}|} \, d\tau = u(a). \tag{6.2}$$

Denote

$$P(\zeta, a) = \frac{1}{2\pi} \frac{1-|a|^2}{|1-\bar{a}e^{i\tau}|^2} = \frac{1}{2\pi} \frac{1-|a|^2}{|\zeta - a|^2},$$

where $\zeta = e^{i\tau}$, and $P(\zeta, a)$ is the **Poisson kernel**. Replacing a by z, (6.2) can be written as

$$\int_0^{2\pi} u(\zeta) P(\zeta, z) \, d\tau = u(z).$$

(6.2) is **Poisson integral formula**. This is the integral representation of the functions which are harmonic on D and continuous on $\overline{D}$. It means that the value of a harmonic function on a point z can be expressed as the integration of the Poisson kernel with the values of the harmonic function on the boundary of the unit disc.

It is easy to extend (6.2) to the following form. Let u be a harmonic function on $D(0, R)$, and continuous on $\overline{D(0, R)}$, then

$$\frac{1}{2\pi} \int_0^{2\pi} u(\zeta) \frac{R^2 - |z|^2}{|\zeta - z|^2} \, d\tau = u(z), \tag{6.3}$$

where $\zeta = Re^{i\tau}$, $z \in D(0, R)$. Similarly,

$$P(\zeta, z) = \frac{1}{2\pi} \frac{R^2 - |z|^2}{|\zeta - z|^2}$$

is Poisson kernel. It is easy to observe that (6.3) can be written as

$$u(z) = \mathrm{Re} \left[\frac{1}{2\pi i} \int_{|\zeta|=R} u(\zeta) \frac{\zeta + z}{\zeta - z} \frac{d\zeta}{\zeta} \right]. \tag{6.4}$$

The function at the bracket on the right side of (6.4) is a holomorphic function on $|z| < R$, thus u is the real part of the holomorphic function

$$f(z) = \frac{1}{2\pi i} \int_{|\zeta|=R} u(\zeta) \frac{\zeta + z}{\zeta - z} \frac{d\zeta}{\zeta} + i\,c, \tag{6.5}$$

where c is an arbitrary real constant. Then $f(z) = u(z) + i\,v(z)$, where

$$v(z) = \frac{1}{2\pi} \int_0^{2\pi} u(\zeta) \frac{2\mathrm{Im}\,(z\overline{\zeta})}{|\zeta - z|^2}\, d\tau + c, \tag{6.6}$$

where $\zeta = Re^{i\tau}$. Of course, $c = v(0)$.

Denote $S(\zeta, z) = \frac{1}{2\pi i} \frac{\zeta + z}{\zeta - z} \frac{1}{\zeta}$, it is the **Schwarz kernel**. (6.5) can be written as

$$f(z) = \int_{|\zeta|=R} u(\zeta) S(\zeta, z)\, d\zeta + i\,v(0).$$

This is another integral representation of holomorphic functions. It means that the value of holomorphic function f at a point $z \in D(0, R)$ can be expressed as the integration of Schwarz kernel $S(\zeta, z)$ with the values of the real part u of f at the boundary $\partial D(0, R)$.

Similary, we denote

$$Q(\zeta, z) = \frac{1}{\pi} \frac{\mathrm{Im}\,(z\overline{\zeta})}{|\zeta - z|^2},$$

it is the **conjugate Poisson kernel**. (6.6) can be written as

$$v(z) = \int_0^{2\pi} u(\zeta) Q(\zeta, z)\, d\zeta + v(0).$$

This is another integral representation of harmonic functions. It means that the value of a harmonic function v at a point $z \in D(0, R)$ can be expressed as the integration of the conjugate Poisson kernel $Q(\zeta, z)$ with the values of the conjugate harmonic function u of v at the boundary $\partial D(0, R)$. Two harmonic functions are conjugate if they are the real part and imaginary part of the same holomorphic function.

Besides the Cauchy kernel, the most important kernel which we discussed above is the Poisson kernel. It has many connetions with other fields. Here we give the connections between the Poisson kernel and two fields as example, one is partial differential equation, one is the harmonic analysis.

Poisson integral and partial differential equation.

In the theory of partial differential equation, one important problem is to find the solution of the Dirichlet problem for partial differential equation of elliptic type. The problem is as follows. To find a function which satisfies the given partial differential equation of elliptic type on a given domain and it coincides with a given function on the boundary of the domain. Using the Poisson integral formula (6.3), we may solve the following **Dirichlet problem**: To find a function which satisfies the Laplace equation on $D(0, R)$ and coincide with a given continuous function $\varphi(Re^{i\tau})$ at the boundary of $D(0, R)$. The solution of this problem is $u(z) = \int_0^{2\pi} P(\zeta, z)\varphi(\zeta)\,d\tau$, where $\zeta = Re^{i\tau}$. The solution is unique.

Now, we prove the above statement.

We express $P(\zeta, z)$ as

$$P(\zeta, z) = \frac{1}{2\pi} \frac{|\zeta|^2 - |z|^2}{|\zeta - z|^2} = \frac{1}{2\pi}\left(\frac{\zeta}{\zeta - z} + \frac{\overline{z}}{\overline{\zeta} - \overline{z}}\right).$$

Obviously $\Delta P = 0$, where $\Delta = 4\frac{\partial}{\partial z}\frac{\partial}{\partial \overline{z}}$ is the Laplace operator. Hence

$$u(z) = \int_0^{2\pi} P(\zeta, z)\varphi(\zeta)\,d\tau$$

satisfies the Laplace equation $\Delta u(z) = 0$ on $D(0, R)$. We will prove $\lim\limits_{z \to \xi} u(z) = \varphi(\xi)$, where $z \in D(0, R), \xi \in \partial D(0, R)$.

We have

$$u(\rho e^{i\theta}) - \varphi(Re^{i\theta_0}) = \int_0^{2\pi} P(\zeta, z)\big(\varphi(\zeta) - \varphi(Re^{i\theta_0})\big)\,d\tau$$

since $\int_0^{2\pi} P(\zeta, z)\,d\tau = 1$, where $\xi = Re^{i\theta_0}, z = \rho e^{i\theta}$ $(0 < \rho < R)$. The assumption is that φ is continuous on $|\zeta| = R$. For any $\varepsilon > 0$, we may choose a $\delta > 0$, such that $|\varphi(Re^{i\theta_0}) - \varphi(Re^{i\tau})| < \varepsilon$ when $|\theta_0 - \tau| < \delta$. Thus

$$u(\rho\, e^{i\theta}) - \varphi(Re^{i\theta_0}) = \left(\int_{|\theta_0 - \tau| < \delta} + \int_{|\theta_0 - \tau| > \delta} \right) P(\zeta, z)\left(\varphi(\zeta) - \varphi(Re^{i\theta_0})\right) d\tau$$
$$= I_1 + I_2.$$

Obviously

$$|I_1| < \varepsilon \int_{|\theta_0 - \tau| < \delta} P(\zeta, z)\, d\tau < \varepsilon.$$

Now we consider I_2. Since $|\theta_0 - \tau| > \delta$, for any given $\varepsilon > 0$ and $\delta > 0$, there exists $\eta > 0$ such that $|\zeta - z| > 2R^2(1 - \cos\delta)$ and $R^2 - \rho^2 < \frac{R^2}{M}(1 - \cos\delta)\varepsilon$ hold when $|z - \xi| < \eta$ and $|\theta_0 - \tau| > \delta$, where $M = \sup_{|\zeta|=R} |\varphi(\zeta)|$. Hence $|I_2| < \varepsilon$.

We have $|u(\rho e^{i\theta}) - \varphi(Re^{i\theta_0})| < 2\varepsilon$ when z is sufficiently closed to ξ. That is $\lim_{z \to \xi} u(\rho e^{i\theta}) = \varphi(Re^{i\theta_0})$.

It is easy to prove the uniqueness. If it is not true, then there exists two solutions u and v, and u and v take the same value $\varphi(Re^{i\theta})$ on the boundary. Hence, $u - v$ takes the value zero on the boundary and $u - v$ is a harmonic function. The only harmonic function which takes the value zero on the boundary is zero function.

Poisson integral and harmonic analysis.

Let $R = 1$ in (6.3), we consider the unit disc D, then the left side of (6.3) is

$$\frac{1}{2\pi} \int_0^{2\pi} u(\zeta) \frac{1 - |z|^2}{|\zeta - z|^2}\, d\tau = \frac{1}{2\pi} \int_0^{2\pi} \frac{(1 - \rho^2)u(e^{i\tau})\, d\tau}{1 - 2\rho\cos(\theta - \tau) + \rho^2} \qquad (6.7)$$

where $z = \rho e^{i\theta}$, $\zeta = e^{i\tau}$.

If $u(e^{i\tau})$ is a given continuous function on ∂D, $u(e^{i\theta})$ has its Fourier series

$$u(e^{i\tau}) \sim \sum_{n=-\infty}^{\infty} a_n e^{in\tau}, \qquad a_n = \frac{1}{2\pi} \int_{-\pi}^{\pi} u(e^{i\tau})e^{-in\tau}\, d\tau.$$

This Fourier series may not convergence. The sum $\sum_{n=-\infty}^{\infty} a_n \rho^{|n|} e^{in\tau}$ is the Abel sum of the Fourier series. By a simple calculation, we obtain that the Abel sum of the Fourier series is the integral

$$\frac{1}{2\pi} \int_0^{2\pi} \frac{(1 - \rho^2)u(e^{i\tau})\, d\tau}{1 - 2\rho\cos(\theta - \tau) + \rho^2}$$

which is the integral on the right side of (6.7). From the solution of the Dirichlet problem as we mentioned above, we know that this integral approaches to $u(e^{i\theta})$ when $\rho \to 1$,

$$\lim_{\rho \to 1} \sum_{n=-\infty}^{\infty} a_n \rho^{|n|} e^{i n\tau} = u(e^{i\tau}).$$

We conclude that if $u(e^{i\tau})$ is a continuous function on the boundary of unit disk, ∂D, its Fourier series is Abel summable to itself. This is an elementary but important theorem in the theory of harmonic analysis.

Finally, we may use the solution of Dirichlet Problem to prove a statement at §2.5: A continuous function with the mean value property is a harmonic function.

Theorem 20 Let $f(z)$ be a continuous real-valued function on a domain $U \subseteq \mathbb{C}$, and let $f(z)$ have the local mean-value property, namely, for any point $z_0 \in U$, there exists a sufficiently small $r_0 > 0$, so that

$$\frac{1}{2\pi} \int_0^{2\pi} f(z_0 + re^{i\theta}) \, d\theta = f(z_0) \qquad (6.8)$$

holds for $0 < r \le r_0$; then $f(z)$ is harmonic on U.

Proof Let z_0 be a point in U, then there exists a $r_0 > 0$, so that (6.8) holds for $0 < r \le r_0$. Let $u_0(\theta) = f(z_0 + r_0 e^{i\theta})$. We may obtain a harmonic function $u(z)$ on $D(z_0, r_0)$ with the boundary value $u_0(\theta)$ by solving a Dirichlet problem. Consider the function $f(z) - u(z)$ on $\overline{D}(z_0, r_0)$. It has the mean value property. In §2.5, we point out that the mean value property implies maximum modulus principle. Thus $|f(z) - u(z)|$ attends the maximum value at $\partial D(z_0, r_0)$. But $f(z) - u(z) = 0$ at the boundary $\partial D(z_0, r_0)$. We have $|f(z) - u(z)| \le 0$, and $f(z) = u(z)$ when $z \in D(z_0, r_0)$ since f and u are continuous functions. Hence, $f(z)$ is harmonic at $z = z_0$ and $f(z)$ is harmonic on U since z_0 is arbitrary in U.

EXERCISES II

1. Use the Cauchy integral formula to evaluate the following integrals:

(i) $\displaystyle \int_{|z+i|=3} \sin z \, \frac{dz}{z+i}$; (ii) $\displaystyle \int_{|z|=2} \frac{e^z}{z-1} \, dz$; (iii) $\displaystyle \int_{|z|=4} = \frac{\cos z}{z^2 - \pi^2} \, dz$;

(iv) $\int_{|z|=4} \dfrac{dz}{(z-1)^n(z-3)}$, $n = 1, 2, \cdots$;

(v) $\int_{|z|=\frac{3}{2}} \dfrac{dz}{(z^2+1)(z^2+4)}$;   (vi) $\int_{|z|=2} \dfrac{dz}{z^5-1}$;

(vii) $\int_{|z|=R} \dfrac{dz}{(z-a)^n(z-b)}$, where a, b are not on the circle $|z| = R$, n is a positive integer;

(viii) $\int_{|z|=3} \dfrac{dz}{(z^3-1)(z-2)^2}$.

2. Show that
$$\left(\frac{z^n}{n!}\right)^2 = \frac{1}{2\pi i}\int_C \frac{z^n e^{z\zeta}}{n!\zeta^n}\cdot\frac{d\zeta}{\zeta},$$
where C is a simple closed curve around origin.

3. If f, g are holomorphic functions on the unit disc $|z| < 1$, and continuous functions on $|z| \le 1$, then
$$\frac{1}{2\pi i}\int_{|\zeta|=1}\left(\frac{f(\zeta)}{\zeta-z} + \frac{zg(\zeta)}{z\zeta-1}\right)d\zeta = \begin{cases} f(z), & \text{when } |z| < 1, \\ g\left(\frac{1}{z}\right), & \text{when } |z| > 1. \end{cases}$$

4. Let
$$f(z) = \int_{|\zeta|=3} \frac{3\zeta^2 + 7\zeta + 1}{\zeta - z}\,d\zeta.$$
Find $f'(1+i)$.

5. Evaluate $\int_{|z|=1}(z + \frac{1}{z})^{2n}\frac{dz}{z}$. Then show that
$$\int_0^{2\pi} \cos^{2n}\theta\,d\theta = 2\pi \cdot \frac{1\cdot 3\cdot 5\cdot\,\cdots\,\cdot(2n-1)}{2\cdot 4\cdot 6\cdot\,\cdots\,\cdot 2n}.$$

6. The Polynomial $p_n(z) = \frac{1}{2^n n!}\frac{d^n}{dz^n}[(z^2-1)^n]$ is the Legendre polynomial. Show that
$$p_n(z) = \frac{1}{2\pi i}\int_\gamma \frac{(\zeta^1-1)^n}{2^n(\zeta-2)^{n+1}}\,d\zeta,$$
where γ is a simple closed curve, z is an inner point of the domain bounded by γ.

7. Suppose $f(z)$ is holomorphic on $\mathbb{C}$ and satisfies the condition $|f(z)| \le Me^{|z|}$, show that $|f(0)| \le M$ and $\frac{|f^{(n)}(0)|}{n!} \le M\left(\frac{e}{n}\right)^n$ $(n = 1, 2, \cdots)$.

8. (Outside Cauchy integral formula) Suppose γ is a rectifiable simple closed curve, inside γ is the domain D_1, outside γ is the domain D_2. If $f(z)$ is holomorphic on D_2, continuous on $D_2 + \gamma$, and $\lim\limits_{z \to \infty} f(z) = A$. Show that

(i) $\dfrac{1}{2\pi i} \displaystyle\int_\gamma \dfrac{f(\zeta)}{\zeta - z} \, d\zeta = \begin{cases} -f(z) + A, & \text{when } z \in D_2, \\ A, & \text{when } z \in D_1; \end{cases}$

(ii) $\dfrac{1}{2\pi i} \displaystyle\int_\gamma \dfrac{z f(\zeta)}{z\zeta - \zeta^2} \, d\zeta = \begin{cases} f(z), & \text{when } z \in D_2, \\ 0, & \text{when } z \in D_1 \end{cases}$

if origin belongs to D_1.

9. Suppose $f(z)$ is holomorphic on a bounded domain D, continuous on the closed domain $\overline{D}$, and $f(z) \neq 0$ on D. Show that if $|f(z)| = M$ (constant) on ∂D, then $f(z) = M e^{i\alpha}$, when α is a real constant.

10. Let $f(z)$ be holomorphic on $\mathbb{C}$, and let a, b be two arbitrary complex constants. Find the limits

$$\lim_{R \to \infty} \int_{|z|=R} \frac{f(z)}{(z-a)(z-b)} \, dz.$$

Then we have another proof of the Liouville theorem.

11. Suppose $f(z)$ is holomorphic on unit disc $|z| < 1$, and

$$|f(z)| \le \frac{1}{1 - |z|}, \qquad |z| < 1.$$

Show that for $0 < r < 1$, we have

$$|f^{(n)}(0)| \le \frac{n!}{r^n (1 - r)}.$$

In particular, let $r = 1 - \frac{1}{n+1}$, we have

$$|f^{(n)}(0)| < e(n + 1)!, \qquad n = 1, 2, \cdots.$$

12. (Integral of Cauchy type) If function $\varphi(\zeta)$ is continuous on a rectifiable curve γ, show that function

$$\Phi(z) = \frac{1}{2\pi i} \int_\gamma \frac{\varphi(\zeta) \, d\zeta}{\zeta - z}$$

is holomorphic on any domain D which does not contain any point of γ. Moreover

$$\Phi^{(n)}(z) = \frac{n!}{2\pi i} \int_\gamma \frac{\varphi(\zeta)\,\mathrm{d}\zeta}{(\zeta - z)^{n+1}}, \qquad n = 1, 2, \cdots$$

hold.

13. Suppose $f(z)$ is holomorphic on the bounded domain D, continuous on $\overline{D}$, and $f(z)$ is not a constant function. Show that if $f(z) \neq 0$ on D, $m = \inf_{z \in \partial D} |f(z)|$, $M = \sup_{z \in \partial D} |f(z)|$, then inequalities $m < |f(z)| < M$ hold for any point z in D.

14. Suppose $p_n(z)$ is a polynomial of degree n, and $|p_n(z)| \leq M$ when $|z| < 1$. Show that $|p_n(z)| \leq M|z|^n$ holds when $1 \leq |z| < +\infty$.

15. Suppose $f(z)$ is holomorphic on disk $|z| < R$, and $|f(z)| \leq M$, $f(0) = 0$. Show that

$$|f(z)| \leq \frac{M}{R}|z|, \qquad |f'(0)| \leq \frac{M}{R}.$$

Equality holds if and only if $f(z) = \frac{M}{R}e^{\mathrm{i}\alpha}z$ where α is a real number.

16. Use Theorem 5 to prove Corollary 1.

17. In the proof of Theorem 13, we used the following conclusion: the order of limit and integration may change if the integrand converges uniformly on any compact subset of domain. Prove this conclusion.

18. Suppose $f(z)$ is holomorphic on $|z| < 1$, and $\operatorname{Re} f(z) > 0$, $f(0) = \alpha > 0$. Show that

$$\left|\frac{f(z) - \alpha}{f(z) + \alpha}\right| \leq |z|, \qquad |f'(0)| \leq 2\alpha$$

when $|z| < 1$.

19. Suppose $f(z)$ is holomorphic on $|z| < 1$, and $f(0) = 0$, $\operatorname{Re} f(z) \leq A$ $(A > 0)$. Show that

$$|f(z)| \leq \frac{2A|z|}{1 - |z|}.$$

20. Find the Taylor expansion and redius of convergence of the following functions at $z = 0$.

(i) $\dfrac{e^z + e^{-z} + 2\cos z}{4}$;   (ii) $\dfrac{z^2 + 4z^4 + z^6}{(1 - z^2)^3}$;

(iii) $(1 - z^{-5})^{-4}$;   (iv) $\dfrac{z^6}{(z^2 - 1)(z + 1)}$.

21. Show that (1) If $f(z)$ is holomorphic on $|z| \leq r$, $f(z) = \sum\limits_{n=0}^{\infty} a_n z^n$ is its Taylor expansion at $z = 0$, then

$$\sum_{n=0}^{\infty} |a_n|^2 r^{2n} = \frac{1}{2\pi} \int_{-\pi}^{\pi} |f(re^{i\theta})|^2 \, d\theta.$$

(2) If $r = 1$ in (1), then

$$\sum_{n=0}^{\infty} \frac{|a_n|^2}{n+1} = \frac{1}{\pi} \iint_{D} |f(z)|^2 \, dA,$$

where D is the unit disk $|z| \leq 1$, dA is the area element.

22. Prove formulas (6.3), (6.4).

23. Show that there exist one root in $|z| < 1$, and three roots in $|z| < 2$ for the equation $z^4 - 6z + 3 = 0$.

24. Find the number of roots in $|z| < 1$ for the equation $z^7 - 5z^4 - z + 2 = 0$.

25. Show that the equation $z^4 + 2z^3 - 2z + 10 = 0$ has one root in each quarter of the plane.

26. Find the number of roots of the equation $z^4 - 8z + 10 = 0$ in $|z| < 1$ and in $1 < |z| < 3$ respectively.

27. Show that the equation $e^z = az^n$ has n roots in $|z| < 1$ if $a > e$.

28. Suppose $f(z)$ is holomorphic on $|z| < 1$, continuous on $|z| \leq 1$ and $|f(z)| < 1$. Show that there exists a unique fixed point on $|z| < 1$.

29. (i) Find a holomorphic function such that its real part is $e^x(x \cos y - y \sin y)$ $(z = x + iy)$.

(ii) Find the most general version of harmonic function in the form $ax^3 + bx^2y + cxy^2 + dy^3$, where a, b, c, d are real constants.

30. Use the mean-value property of harmonic functions to show that

$$\int_0^{\pi} \ln\left(1 - 2r \cos\theta + r^2\right) d\theta = 0$$

if $-1 < r < 1$.

31. Show that if $u(z)$ is a bounded harmonic function on $\mathbb{C}$, then $u(z)$ is identically equal to a constant.

32. Find the harmonic function such that it takes value 1 on an arc on $|z| = 1$ and takes value 0 on the other part of $|z| = 1$.

33. Let U be a domain, and let $f_i(z)$ $(i = 1, 2, \cdots)$ be holomorphic on U, continuous on $\overline{U}$. Show that: if $\sum_{n=1}^{\infty} f_n(z)$ converges uniformly on the boundary ∂U of U, then it converges uniformuly on $\overline{U}$.

34. Suppose $f(z)$ is holomorphic on $D(0, R)$, continuous on $\overline{U}$, and $M = \max_{|z|=R} |f(z)|$. Show that if $z_0 \in D(0, R)\backslash\{0\}$ is a zero point of $f(z)$, then $R|f(0)| \leq (M + |f(0)|)|z_0|$.

35. Suppose $|z_1| > 1, |z_2| > 1, \cdots, |z_n| > 1$. Show that there exists a point z_0 on circle $|z| = 1$, such that $\prod_{k=1}^{n} |z_0 - z_k| > 1$.

36. Suppose $f(z)$ is holomorphic on $D(0, R)$. Show that $M(r) = \max_{|z|=r} |f(z)|$ is an increasing function of r in the interval $[0, R)$.

37. Use the maximum modulus principle to prove the fundementcl theorem of algebra.

38. Suppose $f(z)$ is a non-constant holomorphic function on U, and $f(z)$ has no zero point on U. Show that $|f(z)|$ could not take minimum value on U.

39. (Hadamard three circles theorem) Suppose $0 < r_1 < r_2 < +\infty$, $U = \{z \in \mathbb{C} \mid r_1 < |z| < r_2\}$, $f(z)$ is holomorphic on annulus U, continuous on $\overline{U}$, and $M(r) = \max_{|z|=r} |f(z)|$. Show that $\log M(r)$ is a convex function of $\log r$ on the interval $[r_1, r_2]$. That means the inequality

$$\log M(r) \leq \frac{\log r_2 - \log r}{\log r_2 - \log r_1} \log M(r_1) + \frac{\log r - \log r_1}{\log r_2 - \log r_1} \log M(r_2)$$

holds where $r \in [r_1, r_2]$.

40. Suppose $f(z)$ is holomorphic on $D(0, 1)$ and $f(0) = 0$. Show that the series $\sum_{n=1}^{\infty} f(z^n)$ converges absolutely and uniformly on any compact subset of $D(0, 1)$.

41. Suppose $f(z)$ is holomorphic on $D(0, R)$ and $f(0) = 0$, Show that if the image of $D(0, R)$ by $f(z)$, $f(D(0, R)) \subset D(0, M)$, then

(i) Inequalities $f(z) \leq \frac{M}{R}|z|$, $|f'(0)| = \frac{M}{R}$ hold for all $z \in D(0, R)\backslash\{0\}$.

(ii) Equalities hold in the previous inequalities if and only if $f(z) = \frac{M}{R}e^{i\theta}z$, where θ is any real number.

42. Suppose $f(z)$ is holomorphic on $D(0, 1)$ and $f(0) = 0$. Show that if there exists a constant $A > 0$ such that $\operatorname{Re} f(z) \leq A$ for every $z \in D(0, 1)$, then $|f(z)| \leq \frac{2A|z|}{1-|z|}$ holds for every $z \in D(0, 1)$.

43. Suppose $f(z)$ is holomorphic on $D(0,1)$ and $f(0) = 1$. If for every $z \in D(0,1)$, $\text{Re } f(z) \geq 0$ holds. Use Schwarz lemma to show that

(i) Inequalies

$$\frac{1 - |z|}{1 + |z|} \leq \text{Re } f(z) \leq |f(z)| \leq \frac{1 + |z|}{1 - |z|}$$

hold for every $z \in D(0,1)$.

(ii) Equalities hold in the previous inequalitis when z is non-zero if and only if

$$f(z) = \frac{1 + e^{i\theta} z}{1 - e^{i\theta} z},$$

where θ is any real number.

44. Suppose $f(z)$ is holomorphic on $D(0,1)$. Show that there exists a point $z_0 \in \partial D(0,1)$ and a sequence of points $z_1, z_2, \cdots$ which are inner points of $D(0,1)$ and this sequence converges to z_0, such that $\lim\limits_{n \to \infty} f(z_n)$ exists.

45. Suppose $f(z)$ is holomorphic on $D(0,1)$ and $f(D(0,1)) \subset D(0,1)$. Show that if $z_1, z_2, \cdots, z_n$ are distinct zero points of $f(z)$ on $D(0,1)$, with orders are $k_1, k_2, \cdots, k_n$ respectively, then

$$|f(z)| \leq \prod_{j=1}^{n} \left| \frac{z - z_j}{1 - \overline{z}_j z} \right|^{k_j}$$

holds for every $z \in D(0,1)$. Especially, we have

$$|f(0)| \leq \prod_{j=1}^{n} |z_j|^{k_j}.$$

46. Show that if $f(z)$ is holomophic on $D(0,1)$, and $f(D(0,1)) \subset D(0,1)$, then the inequality

$$M|f'(0)| \leq M^2 - |f(0)|^2$$

holds.

47. Suppose $f(z)$ is holomorphic on $D(0,1)$ and $f(0) = 0$. Show that if $|\text{Re } f(z)| < 1$ holds for every $z \in D(0,1)$, then the following inequalities

(i) $|\text{Re } f(z)| \leq \dfrac{4}{\pi} \text{arctg} |z|$;   (ii) $|\text{Im } f(z)| \leq \dfrac{2}{\pi} \log \left(\dfrac{1 + |z|}{1 - |z|} \right)$

hold for every $z \in D[0,1)$.

48. Find the group of holomorphic automorphism $\operatorname{Aut}(\mathbb{C}^+)$ of the upper half plane $C^+ = \{z \in \mathbb{C} \mid \operatorname{Im} z > 0\}$.

49. Show that if $f(z)$ is holomorphic on $D(0,1) \cup \{1\}$, and $f(0) = 0$, $f(1) = 1$, $f(D(0,1)) \subset D(0,1)$, then $f'(1) \geq 1$.

APPENDIX   Partition of Unity

1. On complex plane, define

$$\theta(z) = \begin{cases} k \exp\left\{\dfrac{-1}{1-|z|^2}\right\}, & \text{when } |z| < 1, \\ 0, & \text{when } |z| \geq 1. \end{cases}$$

where k is a constant, $\int_{\mathbb{C}} \theta(z)\, \mathrm{d}A = 1$, $\mathrm{d}A$ is the area element of $\mathbb{C}$, $\theta(z)$ is strictly greater than zero on $|z| < 1$, and $\operatorname{supp} \theta(z)$ is $|z| \leq 1$. $\theta(z)$ is the standard function on $\mathbb{C}$. If $\varepsilon > 0$ is a constant, set $\theta_\varepsilon(z) = \varepsilon^{-2}\theta\left(\frac{z}{\varepsilon}\right)$, then θ_ε has the same properties of θ, and $\operatorname{supp} \theta_\varepsilon(z)$ is $|z| \leq \varepsilon$.

Let $\Omega \subset \mathbb{C}$ be an open set. All C^∞ real functions on $\mathbb{C}$ with their compact suppports lying in Ω form a space, we denote it as $\mathcal{D}(\Omega)$. We may prove the following result.

2. Let $\Omega \subset \mathbb{C}$ be an open set, B is the basis of open sets of Ω, then there exists a sequence $U_1, U_2, \cdots, U_n, \cdots$. in B, such that

(1) $\underset{j \geq 1}{\cup}\, U_j = \Omega$.

(2) For any compact set K_a in Ω intersects $\{U_j\}_{j \geq 1}$ only finite number of U_j $(j = 1, 2, \cdots)$.

Proof Let $K_{-1} = \phi$, $K_0 = \phi$, $K_1, K_2, \cdots, K_n, \cdots$ be a sequence of compact sets, it exhausts Ω, i.e.,

(i) K_j is contained in the inner part (set of all inner points) $\widetilde{K}_{j+1}$ of K_{j+1};

(ii) $\Omega = \underset{j \geq 1}{\cup} K_j$.

Let $W_r = \widetilde{K}_{r+1} \backslash K_{r-2}$, $V_r = K_r \backslash \widetilde{K}_{r-1}$, $r \geq 1$, then W_r is an open set, V_r is a compact set, and $V_r \subseteq W_r$, $\Omega = \underset{r \geq 1}{\sup} V_r$.

For any point z in V_r, there exists $U_{z,r} \in B$ such that $z \in U_{z,r} \subset W_r$. Since V_r is compact, there exist finite points $z_{r,1}, \cdots, z_{r,k_r}$ in V_r, such that

$$V_r \subseteq \bigcup_{1 \leq i \leq k_r} U_{z_r,i,r} \subseteq W_r.$$

The set $\{z_{z_r,i,r}\}_{r\geq1}$ is countable, and satisfies conditions (1) and (2) due to every compact set K in Ω intersect $\{W_r\}$ finite number.

The set $\{U_j\}_{j\geq1}$ is an open covering of Ω if $\{U_j\}_{j\geq1}$ has property (1). Property (2) means that the open covering is locally finite.

3. Theorem of partition of unity Suppose Ω is a non-empty open subset in $\mathbb{C}$, $\{\Omega_i\}_{i\in I}$ is an open covering of Ω, where I is a set of non-negative integers. Then there exists a sequence $\alpha_1(z), \alpha_2, \cdots, \alpha_n(z), \cdots$ in $\mathcal{D}(\Omega)$ such that

(1) For every $j \geq 1$, there exists a corresponding $i = i(j) \in I$, such that $\operatorname{supp}\alpha_j \subseteq \Omega_i$, and the set $\{\operatorname{supp}\alpha_j\}_{j\geq1}$ is locally finite,

(2) For every $j \geq 1$, $0 \leq \alpha_j \leq 1$,

(3) For every $z \in \Omega$, $\sum_{j\geq1} \alpha_j(z) = 1$.

The sequence $\{\alpha_j(z)\}_{j\geq1}$ is called a C^∞ partition of unity for the covering $\{\Omega_i\}_{i\in I}$.

Proof For every $z \in \Omega$, there exists a $r_z > 0$, such that the closure of the disk centered at z, with raduis r_z, $\overline{B(z, r_z)} \subseteq \Omega_{i_z}$, where $i_z \in I$. All these $B(z,r)$, $z \in \Omega$, $0 < r < r_z$ form a open set basis of Ω. From **2**, there exists a sequence $\{B(z_j, r_j)\}_{j\geq1}$ which satisfies the conditions (1) and (2) in **2**, and

$$B(z_j, r_j) \subseteq \overline{B(z_j, r_j)} \subseteq \Omega_{i(j)},$$

where $i(j) = i_{z_j}$. Let θ be the standerd function, and let $\beta_j(z) = \theta_{r_j}(z - z_j)$, then $\beta_j \in \mathcal{D}(\Omega)$, and $\{\operatorname{supp}\beta_j\}_{j\geq1}$ is locally finite. Hence

$$s(z) = \sum_{j\geq1} \beta_j(z)$$

is C^∞ on Ω, and $s(z) > 0$ when $z \in \Omega$. Let $\alpha_j(z) = \frac{\beta_j(z)}{s(z)}$, then we get the desire sequence in the theorem.

4. If $\Omega \subset \mathbb{C}$ is an open set, K is a compact subset in Ω, V is an open neighborhood of K and $V \subseteq \Omega$, then there exists $\varphi \in \mathcal{D}(V)$ such that

(1) $0 \leq \varphi \leq 1$;

(2) $\varphi \equiv 1$ when z belongs to a neighborhood of K.

Proof For $\varepsilon > 0$, let $V(K, \varepsilon) = \{z \in \mathbb{C} \mid \operatorname{dist}(z, K) < \varepsilon\}$ where $\operatorname{dist}(z, K)$ means the distance between z and K. Choose $\varepsilon > 0$, such that $K \subset V(K, \varepsilon) \subset \overline{V}(K, 2\varepsilon) \subset V$. Let $\Omega_1 = V(K, 2\varepsilon)$, $\Omega_2 = \Omega \backslash \overline{V}(K, \varepsilon)$, then Ω_1, Ω_2 form an open covering of Ω. By the theorem of partition, there exsits

$\{\alpha_j(z)\}_{j\geq 1}$ which satisfies (1), (2), (3) in **3**. Define

$$\varphi(t) = {\sum_j}' \alpha_j(z),$$

where $\sum'$ means that the sum contains the terms which $\operatorname{supp}\alpha_j(z) \subset \Omega_1$ only. Of course, $\varphi(z)$ belongs to $\mathcal{D}(\Omega)$, $\operatorname{supp}\alpha_j(z) \subset V(K, 2\varepsilon)$. If k disappears at $\sum'_j$, then $\operatorname{supp}\alpha_k \not\subset \Omega_1$. Hence $\operatorname{supp}\alpha_k \subset \Omega_2$. Thus $\alpha_k(z) = 0$ on $\overline{V(K,\varepsilon)}$. It implies $\varphi(z) = \sum_{j\geq 1} \alpha_j(z) = 1$ when $z \in V(K,\varepsilon)$. That means $\varphi(z) \equiv 1$ at a neighborhood of K.

CHAPTER III
THEORY OF SERIES OF WEIERSTRASS

§ 3.1   Laurent Series

One important part of Weierstrass theory is to study and to characterize the properties of functions by series . In Chapter I and II, we studied the power series expansion of holomorphic functions and series of functions, etc. Most of these results are the same as the corresponding results in calculus. But, they are different in few places, for example, Theorem 3(3) in Chapter I. Of course, the different parts are more important. In the theory of series, the essential difference between calculus and complex analysis is: besides Taylor series, we have Laurent series in complex analysis. Besides the holomorphic part, the characteristic properties of the functions are decided by the singularity. Laurent series is a powerful tool to study the singularity. Before we introduce Laurent series, we state the Weierstrass theorem of series of functions. It is a deep theorem, and it does not appear in calculus.

Theorem 1 (Weierstrass theorem)   If functions $\{f_n(z)\}$ $(n = 1, 2, \cdots)$ are holomorphic on $U \subset \mathbb{C}$, and the series $\sum\limits_{n=1}^{\infty} f_n(z)$ converges uniformly to $f(z)$ on any compact set in U, then $f(z)$ is holomorphic on U, and $\sum\limits_{n=1}^{\infty} f_n^{(k)}(z)$ converges uniformly to $f^{(k)}(z)$ on any compact set in U $(k = 1, 2, \cdots)$.

Theorem 1 tells us: If a series of holomorphic functions converges uniformly on any compact set in U, then the series converges to a holomorphic function f. Moreover, if we differentiate the series term by term, then the series converges uniformly to the derivative of the holomorphic function f' on any compact set in U. (Recall the theorem of series of functions in calculus. Especially the theorem about the differentiation the series term by term. Then compare it with Theorem 1.)

Proof of Theorem 1   By Theorem 4 of Chapter I, $f(z)$ is defined and continuous on U.

Suppose K is any disk in U, the inner part of K is inside U, γ is any rectifiable closed curve in K. Since $\sum_{n=1}^{\infty} f_n(z)$ converges uniformly on γ, and $f_n(z)$ is holomorphic on K, then

$$\int_{\gamma} f(z)\mathrm{d}z = \sum_{n=1}^{\infty} \int_{\gamma} f_n(z)\mathrm{d}z = 0$$

by Theorem 4 of Chapter I. Since $\int_{\gamma} f(z)\mathrm{d}z = 0$, we have $f(z)$ is holomorphic on K by Morera theorem (Chapter II §2,3, Theorem 7). Thus $f(z)$ is holomorphic on U.

If $z_0 \in U, \overline{D(z_0, r)} \subset U$, then $\sum_{n=1}^{\infty} f_n(\zeta)$ converges uniformly to $f(\zeta)$ on $\partial D(z_0, r)$. When $z \in D\left(z_0, \frac{r}{2}\right)$, we have

$$\sup_{z \in \overline{D(z_0, \frac{r}{2})}} \left| \sum_{j=1}^{n} f_j^{(k)}(z) - f^{(k)}(z) \right| \le c_n \sup_{z \in \overline{D(z_0, r)}} \left| \sum_{j=1}^{n} f_j(z) - f(z) \right|.$$

by Corollary 1 of §2.3 of Chapter II. The right-hand side approaches to zero when $n \to \infty$. Thus $\sum_{j=1}^{\infty} f_j^{(k)}(z)$ converges uniformly to $f^{(k)}(z)$ on $D(z_0, r/2)$. If $\overline{d}$ is any bounded closed domain in U, then $\sum_{j=1}^{\infty} f_j^{(k)}(z)$ converges uniformly to $f^{(k)}(z)$ on the neighborhood of every point on $\overline{d}$. By Heine-Borel theorem, we may select a finite open covering of $\overline{d}$ from them. Thus $\sum_{j=1}^{\infty} f_j^{(k)}(z)$ converges uniformly to $f^{(k)}(z)$ on $\overline{d}$.

Using maximum modulus theorem, we know that: If $D \subset \mathbb{C}$ is a bounded domain, functions $\{f_n(z)\}$ $(n = 1, 2, \cdots)$ are holomorphic on D, and continuous on $\overline{D}$. If $\sum_{n=1}^{\infty} f_n(z)$ converges uniformly on ∂D, then $\sum_{n=1}^{\infty} f_n(z)$ converges uniformly on $\overline{D}$. Thus if we change the condition "$\sum_{n=1}^{\infty} f_n(z)$ converges uniformly to $f(z)$ on any compact set in U" in Theorem 1, to the condition "$\sum_{n=1}^{\infty} f_n(z)$ converges uniformly to $f(z)$ on any closed curve in U", Theorem 1 still holds.

Now we define and discuss the Laurent series.

Let $a \in \mathbb{C}$, c_n $(n = 0, \pm 1, \pm 2, \cdots)$ be complex numbers, then

$$\sum_{n=-\infty}^{\infty} c_n (z - a)^n \tag{1.1}$$

is the Laurent series at point a. The Laurent series is composed by two parts: one part is the power series $\sum_{n=0}^{\infty} c_n(z - a)^n$ with non-negative power terms, another part is the power series $\sum_{n=1}^{\infty} c_{-n}(z - a)^{-n}$ with negative power terms. If these two parts converge at $z = z_0$ both, then the Laurent series converges at this point. If the radius of convergence of $\sum_{n=0}^{\infty} c_n(z - a)^n$ is R, and $R > 0$, then the series converges absolutely on $|z - a| < R$, and converges uniformly on any compact set in $|z - a| < R$. The sum of the series (we denote it by $\varphi(z)$) is holomorphic on $|z - a| < R$. Let $\zeta = \frac{1}{z-a}$, then $\sum_{n=1}^{\infty} c_{-n}(z-a)^{-n} = \sum_{n=1}^{\infty} c_{-n}\zeta^n$. If its radius of convergence is λ, and $\lambda > 0$, then the series converges absolutely on $|\zeta| < \lambda$, and converges uniformly on any compact set in $|\zeta| < \lambda$. Thus the series $\sum_{n=1}^{\infty} c_{-n}(z - a)^{-n}$ converges absolutely on $r = \frac{1}{\lambda} < |z - a| < \infty$, and converges uniformly on any compact set in this interval. The sum of the series (we denote it by $\psi(z)$) is holomorphic on $r < |z - a| < \infty$. If $r > R$, (1.1) diverges everywhere, if $r = R$; (1.1) diverges everywhere except the points on $|z - a| = R$. On $|z - a| = R$, there are different situations. For example, $\sum_{\substack{n=-\infty \\ n \neq 0}}^{\infty} \frac{z^n}{n^2}$ converges everywhere on $|z| = 1$, $\sum_{n=-\infty}^{\infty} z^n$ diverges everywhere on $|z| = 1$; $\sum_{\substack{n=-\infty \\ n \neq 0}}^{\infty} \frac{z^n}{n}$ converges everywhere except the point $z = 1$. If $r < R$, (1.1) converges absolutely on the annulus $r < |z - a| < R$, and converges uniformly on any compact set in this annulus, (1.1) diverges when z is out of the annulus.

This annulus is the convergence annulus. By Theorem 1, the series (1.1) converges to a holomorphic function on the convergence annulus. $\varphi(z)$ is holomorphic on $|z - a| < R$, $\psi(z)$ is holomorphic on $r < |z - a| < \infty$, and $f(z) = \varphi(z) + \psi(z)$ is holomorphic on $r < |z - a| < R$. $\sum_{n=0}^{\infty} c_n(z - a)^n$ is the **holomorphic part** of (1.1), and $\sum_{n=1}^{\infty} c_{-n}(z - a)^{-n}$ is the **principal part** or singularity part of (1.1). The characteristic properties of function $f(z)$ are decided by this part.

We conclude that: If the convergence annulus of Laurent series (1.1) is $r < |z - a| < R$, then (1.1) converges absolutely on this annulus and converges uniformly on any compact set in this annulus, the sum function $f(z)$ is holomorphic on this annulus. Conversely, we have

Theorem 2 If the function $f(z)$ is holomorphic on annulus $V : r < |z - a| < R$ ($0 \leq r < R < \infty$), then $f(z)$ has an expansion

$$f(z) = \sum_{n=-\infty}^{\infty} c_n(z - a)^n \tag{1.2}$$

on V, where

$$c_n = \frac{1}{2\pi i} \int_{|\zeta - a| = \rho} \frac{f(\zeta)\,d\zeta}{(\zeta - a)^{n+1}}, \qquad r < \rho < R. \tag{1.3}$$

The expansion (1.2) is unique. It is the Laurent expansion of $f(z)$, or Laurent series on V.

Proof Obviously, the integrals in (1.3) is independent of ρ ($r < \rho < R$). If $r < \rho_1 < \rho_2 < R$, then

$$\int_{|\zeta - a| = \rho_1} \frac{f(\zeta)\,ds}{(\zeta - a)^{n+1}} = \int_{|\zeta - a| = \rho_2} \frac{f(\zeta)\,d\zeta}{(\zeta - a)^{n+1}}.$$

If $z \in V$, we take $\gamma_1 = \partial D(a, r_1)$, $\gamma_2 = \partial D(a, r_2)$ in V, where $r_1 < r_2$, and z is inside the annulus $r_1 < |z - a| < r_2$ (cf. Figure 3). Using Cauchy integral formula we have

$$f(z) = \frac{1}{2\pi i} \int_{\gamma_2} \frac{f(\zeta)\,d\zeta}{\zeta - z} - \frac{1}{2\pi i} \int_{\gamma_1} \frac{f(\zeta)\,d\zeta}{\zeta - z}. \tag{1.4}$$

When $\zeta \in \gamma_1$, then $\left|\frac{\zeta - a}{z - a}\right| < 1$, and

$$\frac{1}{\zeta - z} = \frac{-1}{(z - a)\left(1 - \frac{\zeta - a}{z - a}\right)} = -\sum_{n=1}^{\infty} \frac{(\zeta - a)^{n-1}}{(z - a)^n}.$$

The series on right-hand side of the previous equality converges uniformly on γ_1. When $\zeta \in \gamma_2$, then $\left|\frac{z - a}{\zeta - a}\right| < 1$, and

$$\frac{1}{\zeta - z} = \frac{1}{(\zeta - a)\left(1 - \frac{z - a}{\zeta - a}\right)} = \sum_{n=0}^{\infty} \frac{(z - a)^n}{(\zeta - z)^{n+1}}.$$

The series on right-hand side of the previous equality converges uniformly on γ_2.

Using these two equalites in (1.4), we have (1.2) and (1.3).

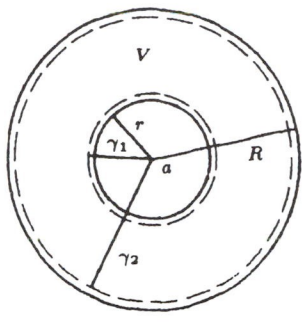

Fig. 3

Finally, we prove the uniqueness. Suppose there is another Laurent expansion

$$f(z) = \sum_{n=-\infty}^{\infty} c'_n(z-a)^n, \qquad r < |z-a| < R. \tag{1.5}$$

The series on the right-hand side of (1.5) converges uniformly to $f(z)$ on $|z - a| = \rho$ $(r < \rho < R)$. Multiplying $\frac{1}{(z-a)^{m+1}}$ on both sides of (1.5), then integrating on $|z - a| = \rho$, we have

$$\int_{|z-a|=\rho} \frac{f(z)\mathrm{d}z}{(z-a)^{m+1}} = \sum_{n=-\infty}^{\infty} c'_n \int_{|z-a|^\rho} (z-a)^{n-m-1}\, \mathrm{d}z = 2\pi\, \mathrm{i}\, c'_m$$

by the uniformly convergence of the series and

$$\int_{|z-a|=\rho} (z-a)^k\, \mathrm{d}z = \begin{cases} 2\pi\mathrm{i}, & \text{if } k = -1, \\ 0, & \text{if } k \neq -1. \end{cases}$$

Thus $c'_m = c_m$ $(m = 0, \pm 1, \pm 2, \cdots)$. We have proved the uniqueness of the series.

§ 3.2 Isolate Singularity

If the function $f(z)$ is holomorphic on a neighborhood $D(a, R)\backslash\{a\}$ of a, the point a is an isolate singularity. By Theorem 2, if a is an isolate singularity of $f(z)$, $f(z)$ can be expanded as a Laurent series

$$f(z) = \sum_{n=-\infty}^{\infty} c_n(z - a)^n,$$

where

$$c_n = \frac{1}{2\pi i} \int_{|\zeta-a|=\rho} \frac{f(\zeta)\,d\zeta}{(\zeta - a)^{n+1}}, \qquad 0 < \rho < R, \quad n = 0, \pm 1, \pm 2, \cdots.$$

As we mentioned on the previous section, $f(z) = \varphi(z) + \psi(z)$, where

$$\varphi(z) = \sum_{n=0}^{\infty} c_n(z - a)^n \tag{2.1}$$

is holomorphic on $|z - a| < R$, it is the holomorphic part of $f(z)$; and

$$\psi(z) = \sum_{n=1}^{\infty} c_{-n}(z - a)^{-n} \tag{2.2}$$

is holomorphic on $0 < |z - a| < \infty$. It is the principal part of $f(z)$.

We consider the limiting value of $\lim_{z \to a} f(z)$. There are three possibilities:

(1) $\lim_{z \to a} f(z)$ exists and is finite. By Riemann theorem (Chapter II §2.3 Theorem 9), $f(z)$ can analytic continuous to $D(a, R)$. (As an exercise, we suggest the reader to prove it directly by Theorem 2). Hence, all c_{-n} are equal to zero in (2.2). Conversely, if all c_{-n} are equal to zero, then $f(z) = \varphi(z)$, and $\lim_{z \to a} f(z) = \varphi(a)$. Thus $\lim_{z \to a} f(z)$ exists and is finite if and only if all c_{-n} are zero. In this situation, we call it as a **removable singularity**.

(2) $\lim_{z \to \infty} f(z)$ exists and is infinite. The necessary and sufficient condition for this situation is that only a finite number of c_{-n} $(n = 1, 2, \cdots)$ are non-zero in (2.2), i,e.,

$$\psi(z) = \frac{c_{-1}}{z - a} + \cdots + \frac{c_{-m}}{(z - a)^m}, \qquad c_{-m} \neq 0,$$

hence

$$f(z) = \varphi(z) + \psi(z)$$

$$= \frac{c_{-m}}{(z-a)^m} + \cdots + \frac{c_{-1}}{z-a} + c_0 + c_1(z-a) + \cdots$$

$$= \frac{g(z)}{(z-a)^m},$$

where
$$g(z) = c_{-m} + c_{-m+1}(z-a) + \cdots, \qquad c_{-m} = g(a) \neq 0.$$

In this situation, we call a as a **pole of order** m of $f(z)$. When $m = 1$, we call it as a **simple pole**.

We prove the above statement. The sufficiency part is obvious. We need to prove the necessary part. Since $\lim_{z \to a} f(z) = \infty$, there exists a $\delta > 0$, so that $f(z) \neq 0$ when $0 < |z - a| < \delta$. Hence $F(z) = \frac{1}{f(z)}$ is holomorphic and non-zero on $0 < |z - a| < \delta$, and $\lim_{z \to a} F(z) = 0$. By (1), a is a removable singularity and it is a zero point of $F(z)$. If a is a zero point of order m of $F(z)$, then $F(z) = (z-a)^m \lambda(z)$ where $\lambda(z)$ is holomorphic on $|z-a| < \delta$, and we may take a neighborhood of a, so that $\lambda(z)$ is non-zero in this neighborhood. Without loss generality, we may assume that the neighborhood is $|z - a| < \delta$. Hence $\frac{1}{\lambda(z)}$ is holomorphic and non-zero on $|z - a| < \delta$, its Taylor expansion is

$$\frac{1}{\lambda(z)} = c_{-m} + c_{-m+1}(z-a) + \cdots, \qquad c_{-m} \neq 0, |z-a| < \delta.$$

Thus

$$f(z) = \frac{1}{F(z)} = \frac{1}{(z-a)^m \lambda(z)} = \frac{c_{-m}}{(z-a)^m} + \cdots + \frac{c_{-1}}{z-a} + c_0 + c_1(z-a) + \cdots.$$

By the uniqueness of Laurent expansion, we have proved it.

From (1) and (2), we have

(3) $\lim_{z \to \infty} f(z)$ does not exist. Then the necessary and sufficient condition for this situation is: there are infinity many non-zero c_{-n}. We call a as the **essential singularity** of $f(z)$. For example, $f(z) = e^{\frac{1}{z}}$, $z = 0$ is an essential singularity because $\lim_{z=x \to 0+} e^{\frac{1}{z}} = +\infty$, $\lim_{z=x \to 0-} e^{\frac{1}{z}} = 0$, and hence $\lim_{z \to 0} f(z)$ does not exist.

For essential singularity, we have the following important theorem.

Theorem 3 (Weierstrass theorem)   If a is an essential singularity of $f(z)$. For any given $\delta > 0$, for any finite complex number A, any $\varepsilon > 0$, there exists a point z in $0 < |z - a| < \delta$ such that $|f(z) - A| < \varepsilon$ holds. It means that in any neighborhood of an essential singularity, the values of $f(z)$ on this neighborhood are dense in $\mathbb{C}$.

Proof   If it is not true, then there exists a finite complex number A and $\varepsilon > 0$, the inequality $|f(z) - A| > \varepsilon$ holds when $0 < |z - a| < \delta$. The function

$$F(z) = \frac{f(z) - A}{z - a}$$

is holomorphic on $0 < |z - a| < \delta$, and $F(z) \to \infty$ when $z \to a$. It implies that a is a pole of $F(z)$. From (2), we have

$$F(z) = \frac{c_{-m}}{(z - a)^m} + \cdots + \frac{c_{-1}}{z - a} + c_0 + c_1(z - a) + \cdots .$$

Thus

$$f(z) = \frac{c_{-m}}{(z - a)^m} + \cdots + \frac{c_{-1}}{z - a} + (A + c_{-1}) + c_0(z - a) + \cdots ,$$

a is a pole of order $m - 1$ of $f(z)$ (when $m > 1$) or a is a removable singularity of $f(z)$ (when $m = 1$). It contradicts with the condition of the theorem.

Weierstrass theorem described the value distribution property of $f(z)$ at an essential singularity. In 1879, Picard proved a more general and more deep theorem. It is the well-known **Picard theorem**. In any neighborhood of an essential singularity, the holomorphic function takes any finite complex number infinity times at most one exceptional number. We will give the proof of Picard theorem in Chapter V.

We have considered the isolate singularity when it is finite complex number in above. Now we consider the isolate singularity when it isthe point at infinity.

Let $f(z)$ be a holomorphic function in annulus: $V : R < |z| < \infty \ (R > 0)$, and let $z = \infty$ be an isolate singularity. Using the transformation $\zeta = \frac{1}{z}$ to transform the neighborhood of $z = \infty$ into the neighborhood of $\zeta = 0$. Then $g(\zeta) = f(z) = f(\frac{1}{\zeta})$ is holomorphic in $0 < |\zeta| < \frac{1}{R}$, and expand it as Laurent series

$$g(\zeta) = \sum_{n=-\infty}^{\infty} c_{-n}\zeta^n = \sum_{n=0}^{\infty} c_{-n}\zeta^n + \sum_{n=1}^{\infty} c_n\zeta^{-n} = \varphi(\zeta) + \psi(\zeta),$$

where $\varphi(\zeta)$ is the holomorphic part of Laurent expansion of $g(\zeta)$, and $\psi(\zeta)$ is the principal part of Laurent expansion of $g(\zeta)$.

Thus

$$f(z) = \sum_{n=-\infty}^{\infty} \frac{c_{-n}}{z^n} = \sum_{n=0}^{\infty} \frac{c_{-n}}{z^n} + \sum_{n=1}^{\infty} c_n z^n = \varphi_0(z) + \psi_0(z),$$

where $\varphi_0(z)$ is the holomorphic part of Laurent expansion of $f(z)$, and $\psi_0(z)$ is the principal part of Laurent expansion of $f(z)$. We have

(1) when $z = \infty$ is a removable singularity, then

$$f(z) = c_0 + \frac{c_{-1}}{z} + \frac{c_{-2}}{z^2} + \cdots ;$$

(2) when $z = \infty$ is a pole of order m, then

$$f(z) = \sum_{n=0}^{\infty} \frac{c_{-n}}{z^n} + c_1 z + c_2 z^2 + \cdots + c_m z^m, \qquad c_m \neq 0;$$

(3) when $z = \infty$ is an essential singularity, then

$$f(z) = \sum_{n=0}^{\infty} \frac{c_{-n}}{z^n} + \sum_{n=1}^{\infty} c_n z^n.$$

§ 3.3   Entire Functions and Meromorphic Functions

A function is an **entire function** if $f(z)$ is holomorphic on $\mathbb{C}$ except the point at infinity. $f(z)$ has the Taylor expansion

$$f(z) = \sum_{n=0}^{\infty} c_n z^n \tag{3.1}$$

on $\mathbb{C}$, and $z = \infty$ is an isolate singularity. By the uniqueness of the Laurent expansion, (3.1) is the Laurent expansion of $f(z)$ at point at infinity. There are three possibilities:

(1) $z = \infty$ is a removable singularity of $f(z)$, then $f(z)$ is a constant by Liouville theorem (Chapter II §2.3 Theorem 8).

(2) $z = \infty$ is a pole of order m of $f(z)$, then $c_n = 0$ when $n > m$. $f(z)$ is a polynomial of degree m,

$$f(z) = c_0 + c_1 z + c_2 z^2 + \cdots + c_m z^m, \qquad c_m \neq 0.$$

(3) $z = \infty$ is an essential singularity of $f(z)$, then

$$f(z) = c_0 + c_1 z + c_2 z^2 + \cdots + c_n z^n + \cdots,$$

there are infinity many non-zero c_n $(n \geq 0)$. $f(z)$ is a **transcendental entire function**, for example, $e^z, \sin z, \cos z$, etc, are transcendental entire functions.

If we do not consider the point at infinity, $f(z)$ has poles only on $\mathbb{C}$ (the number of poles may finite, or may infinity), $f(z)$ is a **meromorphic function**. Entire function is meromorphic function, rational function $f(z) = \frac{P_n(z)}{Q_n(z)}$ is meromorphic function, where $P_n(z), Q_m(z)$ are two polynomials without common factors,

$$P_n(z) = a_0 + a_1 z + \cdots + a_n z^n, \qquad a_n \neq 0,$$
$$Q_n(z) = b_0 + b_1 z + \cdots + b_m z^m, \qquad b_m \neq 0.$$

The zero points of $Q_m(z)$ are poles of $f(z)$, and

$$f(z) = \frac{1}{z^{m-n}} \frac{a_n + \dfrac{a_{n-1}}{z} + \cdots + \dfrac{a_0}{z^n}}{b_m + \dfrac{b_{m-1}}{z} + \cdots + \dfrac{b_0}{z^m}}.$$

Hence

$$\lim_{z \to \infty} f(z) = \begin{cases} \dfrac{a_n}{b_m}, & \text{when } m = n, \\ \infty, & \text{when } n > m, \\ 0, & \text{when } n < m. \end{cases}$$

$z = \infty$ is a removable singularity or a pole of $f(z)$. Conversely, we have the following results.

Theorem 4   If $z = \infty$ is a removable singularity or a pole of $f(z)$, then $f(z)$ is a rational function.

Proof   Since $z = \infty$ is a removable singularity or a pole of $f(z)$, there exists $R > 0$, such that $f(z)$ is holomorphic on $R < |z| < \infty$. If the principal part of the Laurent expansion of $f(z)$ is $p(z)$, then $p(z) \equiv 0$ if $z = \infty$ is a removable singularity of $f(z); p(z)$ is a polynomial if $z = \infty$ is a pole of $f(z)$.

In the disk $|z| \leq R$, $f(z)$ has a finite number of poles. If it is not true, $f(z)$ has infinity many poles in the disk, then by Bolzeno-Weierstrass theorem, these infinity many poles have a limiting point z_0, and z_0 is on $|z| \leq R$. z_0 is a non-isolate singularity. It is impossible, because $f(z)$ is a meromorphic function. Let $z_1, z_2, \cdots, z_k$ be the poles of $f(z)$, and let the principal parts of Laurent expansion of $f(z)$ at the neighborhood of z_i $(i = 1, \cdots, k)$ be

$$\psi_i(z) = \frac{c_{-1}^{(i)}}{z - z_i} + \cdots + \frac{c_{-m}^{(i)}}{(z - z_i)^m}, \qquad i = 1, 2, \cdots, k,$$

and the holomorphic parts be $\varphi_i(z)$. The function

$$F(z) = f(z) - p(z) - \sum_{i=1}^{k} \psi_i(z)$$

is holomorphic on $\mathbb{C}$ except $z_1, z_2, \cdots, z_k, \infty$. But the points $z_1, z_2, \cdots, z_k, \infty$ are removable singularities of $F(z)$. In fact, let $z \to z_i$, $\lim\limits_{z \to z_i} (f(z) - \psi_i(z)) = \varphi_i(z_i)$, and $\sum\limits_{m \neq i} \psi_m(z) - p(z)$ is holomorphic at z_i, thus $\lim\limits_{z \to z_0} F(z)$ exists and is finite. At $z = \infty, f(z) - p(z)$ is the holomorphic part of the Laurent expansion of $f(z)$ at $z = \infty$. Hence $\lim\limits_{z \to \infty} (f(z) - p(z))$ exists and is finite. Obviously, $\lim\limits_{z \to \infty} \sum\limits_{i=1}^{k} \psi_i(z) = 0$. Thus $\lim\limits_{z \to \infty} F(z)$ is finite. It implies $F(z)$ is holomorphic on $\mathbb{C}$. By Liouville theorem, $F(z)$ is a constant c. Finally, we have

$$f(z) = c + p(z) + \sum_{i=1}^{k} \psi_i(z).$$

$f(z)$ is a rational function.

If a meromorphic function is not a rational function, it is a **transcendental meromorphic function.** $z = \infty$ is an essential singularity or $z = \infty$ is the limiting point of poles when the function is a transcendental meromorphic function.

In Chapter II §2.5, we have decided the group of holomorphic automorphisms of the unit disk already (Chapter II Theorem 18). Now we will decide the groups of holomorphic automorphisms of complex plane $\mathbb{C}$ and the extended complex plane $\mathbb{C}^*$, $\mathbb{C}$ added ∞, or equivalently, the Riemann sphere S^2.

We decide the **group of holomorphic automorphisms of complex plane $\mathbb{C}$, Aut $\mathbb{C}$** at first.

If $\alpha(z) \in \text{Aut}(\mathbb{C})$, $\alpha(z)$ maps point at infinity to point at infinity. Since the map is an automorphism, the map is one to one. The point at infinity is a simple pole of $\alpha(z)$. $\alpha(z)$ is a polynomial of degree one, $\alpha(z) = az + b$, $a, b \in \mathbb{C}$, $a \neq 0$. Conversely, it is easy to verify $az + b \in \text{Aut}(\mathbb{C})$. If $a, b \in \mathbb{C}$, $a \neq 0$. Thus Aut $(\mathbb{C})$ is the set of all linear transformations $\{az + b \mid a, b \in \mathbb{C}, a \neq 0\}$, Aut $(\mathbb{C})$ is composed by translations $a(z) = z + b$ and dilations $\alpha(z) = az$.

Then we decide the **group of holomorphic automorphisms of extended complex plane $\mathbb{C}^*$, Aut $(\mathbb{C}^*)$** .

If $\alpha(z) \in \text{Aut}(\mathbb{C}^*)$, and $\alpha(\infty) = \infty$, then in the complex plane $\mathbb{C}, \alpha(z)$ belongs to Aut $(\mathbb{C})$ since automorphism means one to one. Hence $\alpha(z) = cz + d$, where $c, d \in \mathbb{C}$, $c \neq 0$. It is easy to verify that $\alpha(z) = \frac{az+b}{cz+d} \in \text{Aut}(\mathbb{C}^*)$ when $a, b, c, d \in \mathbb{C}$ and $ad - bc \neq 0$.

If $\alpha(z) \in \text{Aut}(\mathbb{C}^*)$, and $\alpha(\infty) \neq \infty$, then $\beta(z) = \frac{1}{\alpha(z) - \alpha(\infty)} \in \text{Aut}(\mathbb{C}^*)$ and $\beta(\infty) = \infty$. Hence, $\beta(z) = cz + d$ where $c, d \in \mathbb{C}$, $c \neq 0$. We have $cz + d = \frac{1}{\alpha(z) - \alpha(\infty)}$. Solving $\alpha(z)$ from this equation, we obtain $\alpha(z) = \frac{az+b}{cz+d}$, where $a = \alpha(\infty)c$, $b = d\alpha(\infty) + 1$. Thus Aut $(\mathbb{C}^*)$ is the set of all linear fractional transformations $\{\frac{az+b}{cz+d} \mid ad - bc = 1\}$. Aut $(\mathbb{C}^*)$ is composed by the translation $\alpha(z) = z + b$, dilation $\alpha(z) = az$ and inversion $\alpha(z) = \frac{1}{z}$. If we establish the one to one correspondence between $\frac{az+b}{cz+d}$ and 2×2 matrix $\left(\begin{smallmatrix} a & b \\ c & d \end{smallmatrix}\right)$, then Aut $(\mathbb{C}^*)$ is isomorphic to the group

$$\left\{ \begin{pmatrix} a & b \\ c & d \end{pmatrix} \,\middle|\, \det \begin{pmatrix} a & b \\ c & d \end{pmatrix} = 1 \right\} / \{\pm I\}$$

where $I = \left(\begin{smallmatrix} 1 & 0 \\ 0 & 1 \end{smallmatrix}\right)$ is the identity matrix, det ( ) means the determinate of the matrix ( ). Actually, this group is $SL(2, \mathbb{C})/\{\pm I\}$, where $SL(2, \mathbb{C})$ is the special linear group of order 2.

In complex analysis, there is a very important theorem.

Uniformization theorem (Poincaré-Koebe theorem)   Any simply connected Riemann surface is one to one holomorphic equivalent to one of the following three domains: the unit disk, the complex plane $\mathbb{C}$, the extended complex plane $\mathbb{C}^*$, i.e., the Riemann sphere S^2.

The meaning of Riemann surface will be given in the next Chapter. The previous results and Theorem 18 of §2.5 of Chapter II decided the group of holomorphic automorphisms of these three domains. By Poincaré-Koebe Theorem, we decided all groups of holomorphic automorphisms of domains which are holomorphic equivalent to all simply connected Riemann surfaces. $\mathbb{C}^*$ has no boundary point, $\mathbb{C}$ has one boundary point. In the next chapter, we will prove the following important Riemann mapping theorem: Any simply connected domain with more than one boundary point is one to one holomorphic equivalent to the unit disk. Up to one to one holomorphic equivalence, we have three simply connected domains only. The geometry of these three domains will be discussed in Chapter V.

The Poincaré-Koebe uniformization theorem is one of the most important and very beautiful theorems in complex analysis.

§ 3.4 Weierstrass Factorization Theorem, Mittag-Leffler Theorem and Interpolation Theorem

We will prove three construction theorems in this section.

For an entire function, it is holomorphic in whole complex plane except the point at infinity. We may express it as a Taylor series (3.1) in §2.3. In §3.3, weknow that if $z = \infty$ is a pole of an entire function, then $f(z)$ is a polynomial. Thus we may regard that an entire function is a natural extension of polynomial, a polynomial (3.1) of degree infinity. For polynomial, an explicit expression is to express it by roots (zero points). If $a_1, \cdots, a_n$ are the roots of polynomial $P_n(z)$, $P_n(z)$ can be expressed as $A(z - a_1) \cdots (z - a_n)$, where A is a complex constant. This expression is called factorization of $P_n(z)$. For a transcendentalentire function, does the factorization of function exist?

If $a_1, a_2, \cdots, a_n, \cdots$ are infinity many zero points of an entire function, can we express this function as $A(z - a_1) \cdots (z - a_n) \cdots = A \prod_{i=1}^{\infty} (z - a_i)$? Since the product is an infinite product, we need to consider the convergence of this product. The correct answer of this question is Weierstrass factorization theorem. We need to discuss infinite product at first.

For a complex number sequence $\{u_n\}$ $(n = 1, 2, \cdots)$, we construct the product

$$p_n = \prod_{k=1}^{n}(1 + u_k).$$

If $1 + u_k \neq 0 \ (k = 1, 2, \cdots)$,

$$\lim_{n \to \infty} p_n = p \neq 0$$

and p is finite, then we say that the infinite product

$$\prod_{n=1}^{\infty}(1 + u_n) \tag{4.1}$$

is **convergent**, and converges to p. We denote it by $p = \prod_{n=1}^{\infty}(1 + u_n)$. Otherwise, (4.1) is **divergent**.When $x \geq 0$, we have $1 + x < e^x$, so

$$|u_1| + |u_2| + \cdots + |u_n| \leq (1 + |u_1|)(1 + |u_2|) \cdots (1 + |u_n|)$$
$$\leq e^{|u_1| + |u_2| + \cdots + |u_n|}.$$

Thus $\sum_{n=1}^{\infty}|u_n|$ and $\prod_{n=1}^{\infty}(1 + |u_n|)$ converge or diverge simultaneously. If $\sum_{n=1}^{\infty}|u_n|$ converges, then we say (4.1) is absolutely convergent. An absolute convergent infinite product is convergent. Moreover, if we rearrange the order of the product, the value of the infinite product is not changed. (We omit the proof of the last statement).

Now we consider the factorization of entire functions.

If the entire function $f(z)$ has no zero point, we may express $f(z) = e^{\varphi(z)}$ where $\varphi(z)$ is an entire function. In fact, we note that $f'(z)/f(z)$ is holomorphic on complex plane, it is the derivative of an entire function $\varphi(z)$. It is easy to verity that the derivative of $f(z)e^{-\varphi(z)}$ is zero. Hence $f(z)$ is $e^{\varphi(z)}$ multiplying a constant. We may put this constant in $\varphi(z)$.

If the entire function $f(z)$ has a finite number of zero points, and $0, a_1, a_2,$ $\cdots, a_n \ (a_i \neq 0, \ i = 1, \cdots, n)$ are the zero points of $f(z)$, its order are $m, m_1,$ $\cdots, m_n$ respectively. Let

$$p(z) = z^m \left(1 - \frac{z}{a_1}\right)^{m_1} \cdots \left(1 - \frac{z}{a_n}\right)^{m_n},$$

then $h(z) = \frac{f(z)}{p(z)}$ has $z = 0$, a_i $(i = 1, \cdots, n)$ as removable singularities. Hence $h(z)$ is an entire function without zero point. So $h(z) = e^{\psi(z)}$, where $\psi(z)$ is an entire function. Thus $f(z)$ can express as

$$f(z) = z^m \left(1 - \frac{z}{a_1}\right)^{m_1} \cdots \left(1 - \frac{z}{a_n}\right)^{m_n} e^{\psi(z)}.$$

That is, $f(z)$ can be expressed as the product of a polynomial and an entire function without zero point. The zero points of the polynomial are the same as the zero points of the function. The order of zero points of the polynomial are the same as the order of the zero points of the function respectively.

If the entire function $f(z)$ has infinite many zero points, and is not identically equal to zero. The zero points of $f(z)$ are enumerable, we may arrange it as a sequence $a_1, a_2, \cdots, a_n, \cdots$ according to the value of the modulus of zero points (except $z = 0$, we will treat it in later),

$$0 < |a_n| \le |a_{n+1}|, \qquad \lim_{n \to \infty} |a_n| = \infty. \tag{4.2}$$

Since $\lim_{n \to \infty} |a_n| = \infty$, for any positive number R, there exists a sequence $k_1, k_2, \cdots, k_n, \cdots$, where k_i $(i = 1, 2, \cdots)$ are positive integers, such that $\sum_{n=1}^{\infty} (\frac{R}{|a_n|})^{k_n+1}$ is convergent. This kind sequence exists, for example, we let $k_n = n - 1$. Consider the infinite product

$$\prod_{n=2}^{\infty} \left(1 - \frac{z}{a_n}\right) \exp\left\{\frac{z}{a_n} + \frac{1}{2}\left(\frac{z}{a_n}\right)^2 + \cdots + \frac{1}{k_n}\left(\frac{z}{a_n}\right)^{k_n}\right\}. \tag{4.3}$$

Let

$$P_n(z) = \frac{z}{a_n} + \frac{1}{2}\left(\frac{z}{a_n}\right)^2 + \cdots + \frac{1}{k_n}\left(\frac{z}{a_n}\right)^{k_n},$$

$$Q_n(z) = \log\left(1 - \frac{z}{a_n}\right) + P_n(z),$$

$$E_n(z) = \left(1 - \frac{z}{a_n}\right)e^{P_n(z)} = e^{Q_n(z)},$$

then (4.3) is $\prod_{n=2}^{\infty} E_n(z)$.

For any fixed positive number R, we may choose a positive integer N, so that $|a_n| > 2R$ when $n \ge N$. Consider the infinite product $\prod_{n=N}^{\infty} E_n(z)$. We have $\frac{|z|}{|a_n|} \leq \frac{1}{2}$, and hence

$$
\begin{aligned}
|Q_n(z)| &\leq \frac{1}{k_n+1}\left(\frac{|z|}{|a_n|}\right)^{k_n+1} + \frac{1}{k_n+2}\left(\frac{|z|}{|a_n|}\right)^{k_n+2} + \cdots \\
&\leq \left(\frac{|z|}{|a_n|}\right)^{k_n+1}\frac{1}{1-\frac{|z|}{|a_n|}} \leq 2\left(\frac{R}{|a_n|}\right)^{k_n+1}
\end{aligned}
$$

when $|z| \leq R$, $n \geq N$. Of course, the series $\sum\limits_{n=1}^{\infty} \left(\frac{R}{|a_n|}\right)^{k_n+1}$ converges. It implies the series $\sum\limits_{n=N}^{\infty} Q_n(z)$ converges absolutely and uniformly on $|z| \leq R$. Consequently,

$$
\prod_{n=N}^{\infty} E_n(z) = \exp\left(\sum_{n=N}^{\infty} Q_n(z)\right)
$$

converges uniformly on $|z| \leq R$. By Weierstrass Theorem (Theorem 1), this infinite product represents a holomorphic function, and it is non-zero. The zero points are located in $|z| \leq 2R$. Hence, all a_n $(n = 1, 2, \cdots)$ which is inside $|z| < R$ are zero points of $\left(1 - \frac{z}{a_1}\right)\prod\limits_{n=1}^{N-1} E_n(z)$, and $f(z)$ has these zero points only in $|z| < R$.

When $|z| < R$, $|Q_n(z)| < 1$ if n is sufficiently large. It is easy to prove that $|e^z - 1| \leq \frac{7}{4}|z|$ when $|z| < 1$. Hence

$$
|E_n(z) - 1| = |e^{Q_n(z)} - 1| \leq \frac{7}{4}|Q_n(z)| \leq \frac{7}{2}\left(\frac{R}{|a_n|}\right)^n.
$$

Thus $\prod\limits_{n=N}^{\infty} E_n(z)$ converges absolutely on $|z| \leq R$ when N is sufficiently large.

We conclude that for a given complex sequence (4.2), there exists an entire function $g(z) = \left(1 - \frac{z}{a_1}\right)\prod\limits_{n=2}^{\infty} E_n(z)$, its zero points are a_n $(n = 1, 2, \cdots)$. This infinite product converges absolutely. Moreover $P_n(z) = \left(1 - \frac{z}{a_1}\right)\prod\limits_{n=2}^{N} E_n(z)$ converges uniformly to $g(z)$ on any disk $|z| < R$.

Theorem 5 (Weierstrass factorization theorem) Suppose $f(z)$ is an entire function, $z = 0$ is the zero point with multiplicity m (m may be zero), $a_1, a_2, \cdots$ $(0 < |a_n| \leq |a_n + 1|, \lim\limits_{n \to \infty} |a_n| = \infty)$ are other zero points of $f(z)$. If for any $R > 0$, there exists a non-negative integer sequence $k_1, k_2, \cdots$,

$k_n, \cdots$, such that $\sum\limits_{n=1}^{\infty} \left(\frac{R}{|a_n|}\right)^{k_n+1}$ converges, then $f(z)$ can be expressed as

$$f(z) = z^m e^{h(z)} \prod_{n=1}^{\infty} \left(1 - \frac{z}{a_n}\right) \exp\left\{\frac{z}{a_n} + \frac{1}{2}\left(\frac{z}{a_n}\right)^2 + \cdots + \frac{1}{k_n}\left(\frac{z}{a_n}\right)^{k_n}\right\}, \quad (4.4)$$

where $h(z)$ is an entire function.

In particular, let $k_n = n - 1$, then

$$f(z) = z^m e^{h(z)} \sum_{n=1}^{\infty} \left(1 - \frac{z}{a_n}\right) \exp\left\{\frac{z}{a_n} + \frac{1}{2}\left(\frac{z}{a_n}\right)^2 + \cdots + \frac{1}{n-1}\left(\frac{z}{a_n}\right)^{n-1}\right\}.$$

Proof As we mentioned already, we may construct an entire function $g(z)$, its zero points are a_n $(n = 1, 2, \cdots)$. Hence $z^m g(z)$ and $f(z)$ has same zero points with same multiplicity at each zero point. Thus $0, a_i$ $(i = 1, 2, \cdots)$ are removable singularity points of $H(z) = \frac{f(z)}{z^m g(z)}$. $H(z)$ has no zero point. We have $H(z) = e^{h(z)}$ where $h(z)$ is an entire function.

Of course $g(z)$ is not unique, so the expression of $f(z)$ is not unique.

For meromorphic function, what is its representation formula? From Weierstrass factorization theorem, we have the following result.

Any meromorphic function can be expressed as a ratio of two entire functions.

Proof If $f(z)$ is a meromorphic function, there exists an entire function $f_1(z)$, its zero points are the poles of $f(z)$. Let $f_2(z) = f(z)f_1(z)$, and define it at the pole a of $f(z)$ as $f_2(a) = \lim\limits_{z \to a} f_2(z)$, then $f_2(z)$ is an entire function. Hence $f(z) = \frac{f_2(z)}{f_1(z)}$.

Any meromorphic function can be expressed as a ratio of two entire function, and every entire function can be expressed as (4.4) by Weierstrass theorem. It means that every meromorphic function can be expressed as a quotient of two expressions of (4.4). The zero points and poles of this meromorphic function are clearly expressed in this form.

Moreover, in last section we know that if a meromorphic function $f(z)$ has a finite number of poles $a_1, a_2, \cdots, a_n$, and $z = \infty$ is its pole or removable singularity, then $f(z)$ is a rational function, and

$$f(z) = c + p(z) + \sum_{j=1}^{n} \psi_j(z),$$

where c is a constant, $p(z)$ is a polynomial, and $\psi_j(z)$ is the principal part of $f(z)$ at $z = a_j$ $(j = 1, \cdots, n)$ (Theorem 4).

For a transcendental meromorphic function $f(z)$, $z = \infty$ is an essential singnlarity or the limiting point of poles. If $z = \infty$ is the essential singularity of $f(z)$, the number of poles of $f(z)$ is finite. Hence $U(z) = f(z) - \sum_{j=1}^{n} \psi_j(z)$ is a transcendental entire function, where $\psi_j(z)$ is the principal part of $f(z)$ at pole $z = a_j$ $(j = 1, 2, \cdots, n)$. Thus

$$f(z) = U(z) + \sum_{j=1}^{n} \psi_j(z).$$

If $z = \infty$ is the limiting point of poles, the poles are $a_1, a_2, \cdots, a_n, \cdots, |a_n| \le |a_{n+1}|$, $\lim_{n \to \infty} |a_n| = \infty$, and the principal part $\psi_j(z)$ at $z = a_j$ $(j = 1, 2, \cdots)$, is given. Does there exist a meromorphic function $f(z)$, its poles are $a_1, a_2, \cdots, a_n, \cdots$, and the corresponding principal parts are $\psi_1, \psi_2, \cdots, \psi_n, \cdots$? The answer is affirmative.

Theorem 6 (Mittag-Leffler theorem) There exists a meromorphic function, its poles are $a_1, a_2, \cdots, a_n, \cdots$, $(|a_n| \le |a_{n+1}|, \lim_{n \to \infty} |a_n| = \infty)$, and its corresponding principal parts are $\psi_1(z), \psi_2(z), \cdots, \psi_n(z), \cdots$.

Proof Let U_i $(i = 1, 2, \cdots)$ be the neighborhood of a_i, and $U_i \cap U_j = \phi$ when $i \ne j$. Let φ_i be a C^∞ function so that $\varphi_i = 1$ at a small neighborhood V_i of a_i, $V_i \subset U_i$, and$\varphi_i = 0$ at the complement of U_i. We define $u = \sum_{i=1}^{\infty} \varphi_i \psi_i$ on $\mathbb{C} \backslash \{a_i\}_1^\infty$, then u is C^∞ at $\mathbb{C} \backslash \{a_i\}_1^\infty$, it equals to ψ_i at $V_i \backslash \{a_i\}$, i.e., at a neighborhood of a_i, the principal value of u is ψ_i. Of course, u is not meromorphic. Let

$$A = \begin{cases} \dfrac{\partial u}{\partial \bar{z}}, & \text{when } z \in \mathbb{C} \backslash \{a_i\}_1^\infty, \\ 0, & \text{when } z = a_i, \quad i = 1, 2, \cdots. \end{cases}$$

Since $u = \psi_i$ when $z \in V_i \backslash \{a_i\}$, we have $\frac{\partial u}{\partial \bar{z}} = 0$ when $z \ne a_i$. Hence $A = 0$ when $z \in V_i \backslash \{a_i\}$. From the definition of A, $A = 0$ when $z = a_i$, thus A is a continuous function. Similarly, we may show that A is a C^∞ function.

By Chapter II §2.1 Theorem 4, the $\bar{\partial}$ equation

$$\frac{\partial v}{\partial \bar{z}} = A$$

has a C^∞ solution v on $\mathbb{C}$, it can express as the formula (1.4) in Chapter

II §2.1. Then $f = u - v$ is the required meromorphic function. Obviously, $\frac{\partial(u-v)}{\partial \bar{z}} = 0$, namely, $\frac{\partial f}{\partial \bar{z}} = 0$ when $z \in \mathbb{C}\backslash\{a_i\}_1^\infty$. Hence f is holomorphic when $z \in \mathbb{C}\backslash\{a_i\}_1^\infty$. By the definition of u, at the neighborhood of $z = a_i$, f has the principal value $\psi_i(z)$ since $v \in C^\infty(\mathbb{C})$.

We have proved the Theorem.

Using the solution of $\bar{\partial}$-equation to prove the Mittag-Leffler theorem, the proof is very simple and beautiful. If we use the classical complex analysis to prove it, we can express the meromorphic function more explcitly.

Theorem 6′ (Mittag-Leffler theorem)   Let $f(z)$ be a meromorphic function, and let a_n $(n = 1, 2, \cdots)$ be the poles of $f(z)$, $|a_n| \leq |a_{n+1}|$, $\lim\limits_{n\to\infty} |a_n| = \infty$, then $f(z)$ can be expressed as

$$f(z) = U(z) + \sum_{n=1}^\infty \{\psi_n(z) - P_n(z)\},$$

where $\psi_n(z)$ is the principal part of $f(z)$ at pole $z = a_n$, $P_n(z)$ is a polynomial $(n = 1, 2, \cdots)$, and $U(z)$ is an entire function.

Proof   Let $\varepsilon_n > 0$ $(n = 1, 2, \cdots)$, and let $\sum\limits_{n=1}^\infty \varepsilon_n$ converge. If $a_1 = 0$, we let $P_1(z) = 0$. For a_n $(\neq 0)$, $\psi_n(z)$ is the polynomial of $\frac{1}{z-a_n}$, and it is holomorphic on $|z| < |a_n|$. We may expand it as a Taylor series

$$\psi_n(z) = \sum_{k=0}^\infty \frac{\psi_n^{(k)}(0)}{k!} z^k$$

when $|z| < |a_n|$. This series converges uniformly to $\psi_n(z)$ on $|z| < \frac{1}{2}|a_n|$. There exists a positive integer λ_n, so that

$$\left| \psi_n(z) - \sum_{k=0}^{\lambda_n} \frac{\psi_n^{(k)}(0)}{k!} z^k \right| < \varepsilon_n.$$

Let $P_n(z) = \sum\limits_{k=0}^{\lambda_n} \frac{\psi_n^{(k)}(0)}{k!} z^k$. Suppose R is any positive number. Let $N = N(R)$ be a positive integer, so that $|a_n| > 2R$ when $n > N$; $|a_n| \leq 2R$ when $n \leq N$, then

$$|\psi_n(z) - P_n(z)| < \varepsilon_n$$

when $n > N$, $|z| < R$ ($|z| < \frac{1}{2}|a_n|$). Since $\sum\limits_{n=1}^{\infty} \varepsilon_n$ converges, $\sum\limits_{n=N+1}^{\infty} \{\psi_n(z) - P_n(z)\}$ converges uniformly on $|z| < R$. When $n > N$, the pole $z = a_n$ is not located in $|z| < R$. By Weierstrass Theorem, $\Phi_N(z) = \sum\limits_{n=N+1}^{\infty} \{\psi_n(z) - P_n(z)\}$ is holomorphic on $|z| < R$. Thus, on $|z| < R$, $\varphi(z) = \sum\limits_{n=1}^{N} \{\psi_n(z) - P_n(z)\} + \Phi_N(z)$ has poles a_n ($n = 1, 2, \cdots$) which satisfy $|a_n| < R$, and the corresponding principal part at a_n is $\psi_n(z)$ ($n = 1, 2, \cdots$). But R is arbitrary, we have that $\varphi(z)$ is a meromorphic function, its poles are $a_1, a_2, \cdots, a_n, \cdots$, and the corresponding principal parts are $\psi_1(z), \psi_2(z), \cdots, \psi_n(z), \cdots$ respectively. Let $U(z) = f(z) - \varphi(z)$, $U(a_n) = \lim\limits_{z \to a_n} \{f(z) - \varphi(z)\}$, then $U(z)$ is an entire function. We have proved the Theorem.

Suppose m points $z_1, \cdots, z_m$ and m complex numbers $a_1, \cdots, a_m$ are given, we may find a polynomial $p(z)$ of degree m such that the value of $p(z)$ at z_j is a_j ($j = 1, 2, \cdots, m$). We only need to solve the coefficients of $p(z)$ from the equations $p(z_j) = a_j$ ($j = 1, 2, \cdots, m$). Similarly, if m points $z_1, \cdots, z_m$, and complex numbers $a_{j,k}$ ($j = 1, 2, \cdots, m$, $0 \le k \le n_j - 1$) are given, where n is a positive integer greater than 1, we may find a polynomial $p(z)$ such that $\frac{p^{(k)}(z_j)}{k!} = a_{j,k}$. That is, we may find a polynomial $p(z)$ such that the first n_j terms of the Taylor expansion of $p(z)$ at z_j is the given polynomial. We have the following very general interpolation theorem.

Theorem 7 (Interpolation theorem) Let $z_1, z_2, \cdots$ be a discrete point set in $\mathbb{C}$. Let $n_1, n_2, \cdots$ be a positive integer sequence, and $a_{j,k}$ ($j \ge 1$, $0 \le k \le n_j - 1$) be a sequence of complex numbers, then there exists an entire function $g(z)$ such that

$$g^{(k)}(z_j) = k! a_{j,k}, \qquad j \ge 1, \quad 0 \le k \le n_j - 1.$$

That means, if a sequence of points $\{z_j\}$ and the first n_j terms of the Taylor expansion at z_j ($j = 1, 2, \cdots$) are given, then there exists an entire function, the first n_j terms of the Taylor expansion of this function at z_j ($j = 1, 2, \cdots$) are coincided with the given polynomials

Proof Using Theorem 5 (Weierstrass factorization theorem), we may find an entire function $f(z)$, such that $f(z)$ has zero points z_j with multiplicity n_j ($j = 1, 2, \cdots$). Since $z_1, z_2, \cdots, z_n, \cdots$ is a discrete sequence, we may choose a positive number sequence $\varepsilon_1, \varepsilon_2, \cdots, \varepsilon_n$, such that all the disks which is centered at z_j, with radius $2\varepsilon_j$ $(j = 1, 2, \cdots)$ are mutually non-intersected. Let

$$P_j(z) = \sum_{0 \leq k \leq n_j - 1} a_{j,k}(z - z_j)^k, \qquad j \geq 1.$$

For every j, we have $\varphi_j \in C^\infty$, so that the support of φ_j is inside $D(z_j, 2\varepsilon_j)$, $0 \leq \varphi_j \leq 1$, and $\varphi_j \equiv 1$ on $\overline{D}(z_j, \varepsilon_j)$ $(j = 1, 2, \cdots)$.

Let $\psi(z) \in C^\infty$, and let

$$g(z) = \sum_{j \geq 1} P_j(z)\varphi_j(z) - f(z)\psi(z). \tag{4.5}$$

Since the supports of φ_j $(j = 1, 2, \cdots)$ are mutually non-intersected, at most one term in $\sum_{j \geq 1} P_j(z)\varphi_j(z)$ is non-zero at every point $z \in \mathbb{C}$. Thus the sum $\sum_{j \geq 1} P_j(z)\varphi_j(z)$ makes sense. We try to find ψ to make $g(z)$ become an entire function. That means, we need $\frac{\partial g}{\partial \bar{z}} = 0$. It is

$$\sum_{j \geq 1} P_j(z)\frac{\partial \varphi_j(z)}{\partial \bar{z}} = f(z)\frac{\partial \psi}{\partial \bar{z}}$$

for any $z \in \mathbb{C}$. Let $h(z) = \sum_{j \geq 1} p_j(z)\frac{\partial \varphi_j(z)}{\partial \bar{z}}$, then $h(z) \equiv 0$ on $\bigcup_j \overline{D(z_j, \varepsilon_j)}$. Let $\frac{h(z)}{f(z)} = 0$ at $z = z_j$, then $\frac{h(z)}{f(z)}$ is C^∞ on $\mathbb{C}$. Since h has a mutually non-intersect compact support, the $\bar{\partial}$-equation $\frac{\partial \psi}{\partial \bar{z}} = \frac{h}{f}$ has a C^∞ solution ψ by Theorem 4 of §2.1 Chapter II. Taking this ψ in (4.5), (4.5) defines an entire function, and $\frac{\partial \psi}{\partial \bar{z}} = 0$ on $\bigcup_j \overline{D(z_j, \varepsilon_j)}$. Thus ψ is holomorphic on a neighborhood of z_j. It can be verified directly, that is

$$g^{(k)}(z_j) = P_j^k(z_j) = k!a_{jk}, \qquad 0 \leq k \leq n_j - 1,$$

since $f(z)$ has a zero point with multiplicity n_j.

We have proved the theorem.

If a power series is given, in general, we can not find an entire function whose Taylor series is the given power series. Theorem 7 tells us that for any polynomial of degree n, even n is very large (the important point is that n is finite), we can find an entire function, its first n terms section is the given polynomial. It indicates the essential difference between polynomial and entire function. It shows that Theorem 7 is a deep theorem.

§ 3.5 Residue Theorem

Let $f(z)$ be a holomorphic function on $D(a,r)\backslash\{a\}$ $(r > 0)$, and let a be an isolate singularity of $f(z)$. The residue of $f(z)$ at a is defined as

$$\text{Res}(f,a) = \frac{1}{2\pi i} \int_{|z-a|=\rho} f(z)\,dz,$$

where $0 < \rho < r$. We may expand $f(z)$ as a Laurent series $f(z) = \sum_{n=-\infty}^{\infty} c_n(z-a)^n$ on $D(a,r)\backslash\{a\}$. The residue of $f(z)$ at a, $\text{Res}(f,a)$, is c_{-1}.

If $z = \infty$ is an isolate singularity of $f(z)$, $f(z)$ is holomorphic on $R < |z| < \infty$. We define the residue of $f(z)$ at $z = \infty$ as

$$\text{Res}(f,\infty) = \frac{-1}{2\pi i} \int_{|z|=\rho} f(z)\,dz$$

where $R < \rho < \infty$. We may expand $f(z)$ as a Laurent series $f(z) = \sum_{z=-\infty}^{\infty} c_n z^n$ in the neighborhood of $z = \infty$. The residue of $f(z)$ at $z = \infty$, $\text{Res}(f,\infty)$, is $-c_{-1}$.

If a $(\neq \infty)$ is a pole of order m of $f(z)$, we may express $f(z)$ at a neighborhood of a by

$$f(z) = \frac{1}{(z-a)^n} g(z),$$

where $g(z)$ is holomorphic at $z = a$, and $g(z) \neq 0$. Hence

$$g(z) = \sum_{a=0}^{\infty} \frac{1}{n!} g^{(n)}(a)(z-a)^n.$$

Thus

$$\text{Res}(f,a) = c_{-1} = \frac{1}{(m-1)!} g^{(m-1)}(a).$$

Since

$$g^{(m-1)}(a) = \lim_{z \to a} \frac{d^{m-1}}{dz^{m-1}} \{(z-a)^m f(z)\},$$

we have

$$\text{Res}(f,a) = \frac{1}{(m-1)!} \frac{d^{m-1}}{dz^{m-1}} \{(z-a)^m f(z)\}.$$

In particular, when $m = 1$, we have

$$\operatorname{Res}(f, a) = g(z) = \lim_{z \to \infty} (z - a)f(z).$$

Theorem 8 (Residue theorem)   If $f(z)$ is a holomorphic function on $U \subset \mathbb{C}$ except the points $z_1, z_2, \cdots, z_n$, $f(z)$ is continuous on $\overline{U}$ except the points $z_1, z_2, \cdots, z_n$, and ∂U is a rectifiable closed curve, then

$$\int_{\partial U} f(z)\,\mathrm{d}z = 2\pi i \sum_{R=1}^{n} \operatorname{Res}(f, z_k).$$

Theorem 8′ (Residue theorem)   If $f(z)$ is a holomorphic function on $\mathbb{C}^*$ except the points $z_1, z_2, \cdots, z_n, \infty$, and $z_1, z_2, \cdots, z_n, \infty$ are isolate singularities of $f(z)$, then the sum of residues of $f(z)$ at all these points is equal to zero,

$$\sum_{R=1}^{n} \operatorname{Res}(f, z_k) + \operatorname{Res}(f, \infty) = 0.$$

Using Cauchy integral theorem, it is easy to prove these two theorems. We omit the proofs.

The residue theorems themselves are simple, the important point is to use them to evaluate the value of some definite integers. Usually, we can not find the primitive function of the integral, but we may use the residue theorem to find the value of the definite integral if we choose carefully the complex function $f(z)$, the contour of integration, etc. Here we give three simple examples only.

Example 1   Evaluate the integral $\int\limits_{-\infty}^{\infty} \frac{\mathrm{d}x}{(1+x^2)^{n+1}}$, where n is a positive integer.

Solution   Let $f(z) = \frac{1}{(1+z^2)^{n+1}}$, then $f(z)$ has an unique isolate singularity $z = i$ on the upper half plane. It is the pole of order $n+1$ of $f(z)$. Let U be the upper semi-disk $|z| < R$, $\operatorname{Im} z > 0$ (Figure 4), then

$$
\begin{aligned}
\operatorname{Res}(f, i) &= \frac{1}{n!} \frac{\mathrm{d}^n}{\mathrm{d}z^n} \left\{ \frac{(z-i)^{n+1}}{(z^2+1)^{n+1}} \right\}\Big|_{z=i} = \frac{1}{n!} \frac{\mathrm{d}^n}{\mathrm{d}z^n} \left\{ \frac{1}{(z+i)^{n+1}} \right\}\Big|_{z=i} \\
&= \frac{1}{n!} \frac{(-1)^n (n+1)(n+2)\cdots(2n)}{(2i)^{2n+1}} = \frac{1}{2i} \frac{(2n)!}{2^{2n}(n!)^2}.
\end{aligned}
$$

On the other hand,

$$
\begin{aligned}
\operatorname{Res}(f,\mathrm{i}) &= \frac{1}{2\pi\mathrm{i}} \int_{\partial U} \frac{\mathrm{d}z}{(1+z^2)^{n+1}} \\
&= \frac{1}{2\pi\mathrm{i}} \int_{-R}^{R} \frac{\mathrm{d}x}{(1+x^2)^{n+1}} + \frac{1}{2\pi\mathrm{i}} \int_{\gamma_R} \frac{\mathrm{d}z}{(1+z^2)^{n+1}} \qquad (5.1)
\end{aligned}
$$

where γ_R is the upper semi-circle (Figure 4): $z = Re^{\mathrm{i}\theta}$ ($0 \le \theta \le \pi$). Obviously,

$$
\frac{1}{2\pi\mathrm{i}} \int_{\gamma_R} \frac{\mathrm{d}z}{(1+z^2)^{n+1}} = \frac{1}{2\pi\mathrm{i}} \int_{0}^{\pi} aRe^{\mathrm{i}\theta} \frac{\mathrm{d}\theta}{(1+R^2 e^{2\mathrm{i}\theta})^{n+1}},
$$

the integral$\to 0$ when $R \to \infty$. Let $R \to \infty$ in (5.1), we have

$$
\int_{-\infty}^{\infty} \frac{\mathrm{d}x}{(1+x^2)^{n+1}} = 2\pi\,\mathrm{i}\,\operatorname{Res}(f,\mathrm{i}) = \pi \frac{(2n)!}{2^{2n}(n!)^2}.
$$

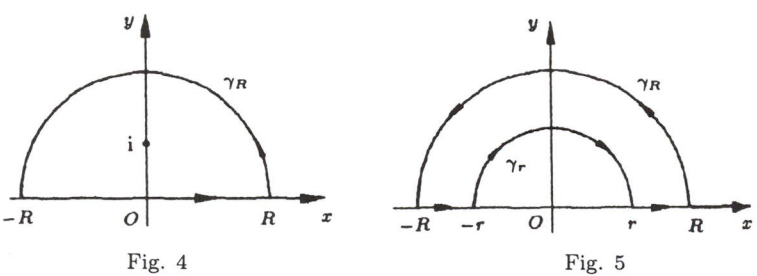

Fig. 4   Fig. 5

Example 2   Evaluate the integral (Dirichlet integral) $\int_0^\infty \frac{\sin x}{x}\,\mathrm{d}x$.

Solution   Obviously $\int_0^\infty \frac{\sin x}{x}\,\mathrm{d}x = \frac{1}{2}\int_{-\infty}^{\infty} \frac{\sin x}{x}\,\mathrm{d}x$. Let $f(z) = \frac{e^{\mathrm{i}z}}{z}$, and let U be the semi-annulus on the upper half-plane (Figure 5), then its boundary contains four parts:

$$
\begin{aligned}
-R &< z < -r; \qquad r < z < R; \\
\gamma_r &: z = re^{\mathrm{i}\theta}, \qquad 0 \le \theta \le \pi; \\
\gamma_R &: z = Re^{\mathrm{i}\theta}, \qquad 0 \le \theta \le \pi.
\end{aligned}
$$

$f(z)$ is holomorphic on this semi-annulus,

$$
\int_r^R f(x)\,\mathrm{d}x + \int_{-R}^{-r} f(x)\,\mathrm{d}x + \int_{\gamma_R} f(z)\,\mathrm{d}z + \int_{\gamma_r} f(z)\,\mathrm{d}z = 0
$$

by Cauchy integral theorem. Since

$$\int_{\gamma_R} f(z)\,\mathrm{d}z = \int_0^\pi \frac{e^{\mathrm{i}\,R(\cos\theta + \mathrm{i}\,\sin\theta)}}{Re^{\mathrm{i}\theta}}\,\mathrm{i}Re^{\mathrm{i}\theta}\,\mathrm{d}\theta = \mathrm{i}\int_0^R e^{-R\sin\theta + \mathrm{i}\,R\cos\theta}\,\mathrm{d}\theta,$$

we have

$$\left|\int_{\gamma_R} f(z)\,\mathrm{d}z\right| \le \int_0^\pi e^{-R\sin\theta}\,\mathrm{d}\theta = 2\int_0^{\frac{\pi}{2}} e^{-R\sin\theta}\,\mathrm{d}\theta.$$

The inequality $\frac{2}{\pi}\theta \le \sin\theta$ holds when $0 \le \theta \le \frac{\pi}{2}$, the integral

$$\int_0^{\frac{\pi}{2}} e^{-R\sin\theta}\,\mathrm{d}\theta \le \int_0^{\frac{\pi}{2}} e^{-\frac{2R}{\pi}\theta}\,\mathrm{d}\theta = \frac{\pi}{2R}(1 - e^{-R}).$$

It approaches to zero when $R \to \infty$. Hence $\int_{\gamma_R} f(z)\mathrm{d}z \to 0$ when $R \to \infty$. On the other hand, the integral

$$\int_{\gamma_r} f(z)\,\mathrm{d}z = \mathrm{i}\int_\pi^0 e^{-r\sin\theta + \mathrm{i}\,r\cos\theta}\,\mathrm{d}\theta$$

$$= \mathrm{i}\int_\pi^0 \left(1 + O(r)\right)\mathrm{d}\theta = -\pi\mathrm{i} + O(r).$$

$\int_{\gamma_r} f(z)\,\mathrm{d}z \to -\pi\,\mathrm{i}$ when $r \to 0$. The integral

$$\int_{-R}^{-r} f(x)\,\mathrm{d}x = \int_{-R}^{-r} \frac{e^{\mathrm{i}x}}{x}\,\mathrm{d}x = -\int_r^R \frac{e^{-\mathrm{i}x}}{x}\,\mathrm{d}x.$$

Let $r \to 0$, $R \to \infty$ in the following equality

$$\int_r^R \frac{e^{\mathrm{i}x} - e^{-\mathrm{i}x}}{x}\,\mathrm{d}x + \int_{\gamma_R} f(z)\,\mathrm{d}z + \int_{\gamma_r} f(z)\,\mathrm{d}z = 0,$$

we have

$$\int_0^\infty \frac{\sin x}{x}\,\mathrm{d}x = \frac{\pi}{2}.$$

Example 3   Evaluate the integrals $\int_0^\infty \cos x^2\,\mathrm{d}x$ and $\int_0^\infty \sin x^2\,\mathrm{d}x$.

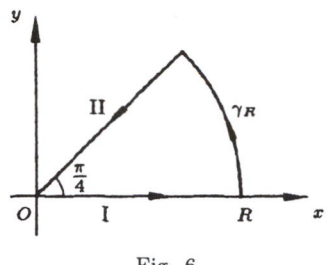

Fig. 6

Solution Let $f(z) = e^{\mathrm{i}z^2}, U$ be a sector diagram bounded by I: $0 \leq z \leq R$; II: $re^{\mathrm{i}\frac{\pi}{4}}$ $(0 \leq r \leq R)$, and $\gamma_R : Re^{\mathrm{i}\theta}$ $(0 \leq \theta \leq \frac{\pi}{4})$ (Figure 6). By Cauchy integral theorem,

$$\int_{\mathrm{I}} f(z)\,\mathrm{d}z + \int_{\mathrm{II}} f(z)\,\mathrm{d}z + \int_{\gamma_R} f(z)\,\mathrm{d}z = 0.$$

The integral

$$\int_{\gamma_R} f(z)\,\mathrm{d}z = \int_0^{\frac{\pi}{4}} e^{\mathrm{i}\,R^2(\cos 2\theta + \mathrm{i}\,\sin 2\theta)}\mathrm{i}\,Re^{\mathrm{i}\theta}\,\mathrm{d}\theta,$$

the modulus of the integral

$$\left|\int_{\gamma_R} f(z)\,\mathrm{d}z\right| \leq R \int_0^{\frac{\pi}{4}} e^{-R^2 \sin 2\theta}\,\mathrm{d}\theta \leq R \int_0^{\frac{\pi}{4}} e^{-R^2 \frac{4}{\pi}\theta}\,\mathrm{d}\theta = \frac{\pi}{4R}(1 - e^{-R^2}).$$

The right-hand side of the previous inequality approaches to zero when $R \to \infty$. We have

$$\lim_{R \to \infty} \left(\int_{\mathrm{I}} f(z)\,\mathrm{d}z + \int_{\mathrm{II}} f(z)\,\mathrm{d}z\right) = 0.$$

It is,

$$\int_0^\infty e^{\mathrm{i}\,x^2}\,\mathrm{d}x - \int_0^\infty e^{\mathrm{i}\,x^2 e^{\mathrm{i}\frac{\pi}{2}}}\,e^{\frac{\pi\mathrm{i}}{4}}\,\mathrm{d}x = 0.$$

We have

$$\int_0^\infty e^{\mathrm{i}\,x^2}\,\mathrm{d}x = e^{\frac{\pi\mathrm{i}}{4}} \int_0^\infty e^{-x^2}\,\mathrm{d}x.$$

We have known that $\int_0^\infty e^{-x^2}\,\mathrm{d}x = \frac{\sqrt{\pi}}{2}$ already, hence

$$\int_0^\infty e^{ix^2}\, dx = \frac{\sqrt{\pi}}{2} e^{\frac{\pi i}{4}}.$$

Thus
$$\int_0^\infty \cos x^2\, dx = \int_0^\infty \sin x^2\, dx = \frac{\sqrt{2\pi}}{4}.$$

§ 3.6   Analytic Continuation

Let $f(z)$ be a holomorphic function on domain $U \subset \mathbb{C}$. If there exists a domain G which contains U, and a function $F(z)$ which is holomorphic on G, such that $F(z) = f(z)$ on U, then we say that $F(z)$ is an analytic continuation (or holomorphic continuation) of f from U to G. By the uniqueness of holomophic function, if F exists on G, it is unique. Similarly, if $f_1(z), f_2(z)$ are holomorphic on U_1, U_2 respectively, and $U_1 \cap U_2 = U_3 \neq \phi$, $f_1 = f_2$ on U_3, we may define $f(z)$ on $U = U_1 \cup U_2$ (Figure 7) by

$$f(z) = \begin{cases} f_1(z), & z \in U_1, \\ f_2(z), & z \in U_2. \end{cases}$$

Thus f is holomorphic on U, we call that f_1 is an analytic continuation of f_2, and f_2 is an analytic continuation of f_1.

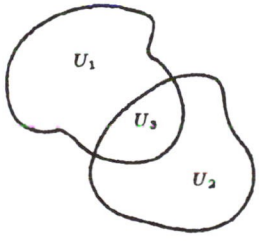

Fig. 7

The most natural and most important method of analytic continuation is using power series. By Abel theorem (Chapter I, §1.6 Theorem 3), there exists a radius of convergence of a power series

$$a_0 + a_1 z + \cdots + a_n z^n + \cdots. \tag{6.1}$$

The series converges absolutely on $|z| < R$ and converges uniformly on any compact set in $|z| < R$; it is a holomorphic function, we denote it by $f(z)$. If $z_0 \in D(0, R)$, we may expand it as a Taylor series

$$f(z) = \sum_{n=0}^{\infty} \frac{f^n(z_0)}{n!} (z - z_0)^n$$

at $z = z_0$. If its radius of convergence is ρ, then $\rho \geq R - |z_0|$. If $\rho > R - |z_0|$, then some part of $D(z_0, \rho)$ is outside $D(0, R)$. $f(z)$ can be analytically continued to $D(z_0, \rho) \backslash D(0, R)$. If $\rho = R - |z_0|$, then $D(z_0, \rho)$ and $D(0, R)$ are tangent. If the tangent point is ζ_0, it means that $f(z)$ can not be analytically continuaed at ζ_0, ζ_0 is a singularity of $f(z)$. If R is the radius of convergence of (6.1), there exists one singularity of $f(z)$ on $|z| = R$ at least. Otherwise, $f(z)$ can be analytically continued at any point on $|z| = R$. That means for any point ζ on $|z| = R$, there exist $D(\zeta, \gamma_\zeta)$ and $g_\zeta(z), g_\zeta(z)$ is holomorphic on $D(\zeta, \gamma_\zeta)$ and $g_\zeta(z) = f(z)$ when $z \in D(0, R) \cap D(\zeta, \gamma_\zeta)$. Since $|z| = R$ is compact, by Heine-Borel theorem, we may select a finite number of $D(\zeta_1, r_{\zeta_1}), D(\zeta_2, r_{\zeta_2}), \cdots, D(\zeta_m, r_{\zeta_m})$ from $\{D(\zeta, r_\zeta)\}$ such that it covers $|z| = R$. Let $G = \bigcup_{k=1}^{m} D(\zeta_k, \gamma_{\zeta_k})$, ρ be the distance between $|z| = R$ and ∂G, then $\rho > 0$, and $\{R - \rho < |z| < R + \rho\} \subset G$. In G, we define $\Phi(z) = g_{\zeta_k}(z)$ when $z \in D(\zeta_k, r_{\zeta_k})(k = 1, 2, \cdots, m)$, then $\Phi(z)$ is a single-valued holomorphic function. If $D(\zeta_k, r_{\zeta_k}) \cap D(\zeta_l, r_{\zeta_l}) \neq \phi$, $k \neq l$, then $D(\zeta_k, r_{\zeta_k}) \cap D(\zeta_l, r_{\zeta_l}) \cap D(0, R) \neq \phi$. In this part, $g_{\zeta_k}(z) = g_{\zeta_l}(z) = f(z)$. By the uniqueness theorem of holomorphic function, $g_{\zeta_k}(z) = g_{\zeta_l}(z)$ on $D(\zeta_k, r_{\zeta_k}) \cap D(\zeta_l, r_{\zeta_l})$, and $\Phi(z) = f(z)$ on $G \cap D(0, R)$. Hence $f(z)$ can be analytically continued to $G \cup D$ which contains $D(0, R + \rho)$. It contradicts with the radius of convergence of (6.1).

Thus there exists at least one singularity on the convergence circle. The following example is a well-known example, it shows that every point on the convergence circle is a singularity.

Example $f(z) = z^{1!} + z^{2!} + \cdots + z^{n!} + \cdots .$ (6.2)

We know that

$$\varlimsup_{n \to \infty} \sqrt[n]{|a_n|} = 1$$

since

$$a_n = \begin{cases} 1, & \text{when } n = k!, \\ 0, & \text{when } n \neq k!. \end{cases}$$

The radius of convergence of (6.2) is $R = 1$, $f(z)$ is holomorphic on $D(0, 1)$. If $z_0 \in D(0, 1)$ and $|z_0| = \frac{1}{2}$, the Taylor expansion of $f(z)$ at $z = z_0$ is

$$g(z) = \sum_{n=0}^{\infty} \frac{f^{(n)}(z_0)}{n!} (z - z_0)^n.$$

Extend the line segment Oz_0 and intersect $|z| = 1$ at ζ_0. If we can prove that the radius of convergence of $g(z)$ is $\frac{1}{2}$, then it means that $f(z)$ cannot be analytically continued at ζ_0. If it is not true, the radius of convergence of $g(z)$ is $\rho > \frac{1}{2}$, then $D(\zeta_0, \rho) \cap D(0, 1) \neq \phi$. There exists a circle arc σ on $|z| = 1$ which is inside $D(\zeta_0, \rho)$, and $\zeta_0 \in \sigma$. The points $\exp\{\frac{2\pi i p}{q}\}$ (p, q are integers, $\frac{p}{q}$ is an irreducible fraction) are everywhere dense on $|z| = 1$. On σ there exists a point $\zeta_1 = \exp\{\frac{2\pi i p}{q}\}$ and $\lim_{r \to 1} g(r\zeta_1) = g(\zeta_1)$ $(0 < r < 1)$. Of Course, $g(z) = f(z)$ when $z \in D(0, 1)$, hence $\lim_{r \to 1} f(r\zeta_1) = g(\zeta_1)$. Since

$$f(r\zeta_1) = \sum_{n=1}^{q-1} r^{n!}\zeta_1^{n!} + \sum_{n=1}^{\infty} r^{n!},$$

and

$$\sum_{n=1}^{\infty} r^{n!} > \sum_{n=1}^{N} r^{n!} > (N - q)r^{N!},$$

the sum $\sum_{n=q}^{\infty} r^{n!}$ may greater than any positive integer, when $r \to 1$. Hence $\lim_{r \to 1} |f(r\zeta_1)| = \infty$. It contradicts with $\lim_{r \to 1} f(r\zeta_1) = g(\zeta_1)$. Thus ζ_0 is a singularity of $f(z)$.

A function is analytic (or holomorphic) at a neighborhood of $z = z_0$ if it can be expanded as a convergent power series at the neighborhood of $z = z_0$. This is the definition of a local analytic (or local holomorphic) function. This definition is consistent with the definition of an analytic (or holomorphic) function in Chapter I §1.3. Now we may define the global analytic (or holomorphic) function when we use the idea of analytic continuation.

We start from a local holomorphic function $f(z) = \sum_{n=0}^{\infty} c_n(z - a)^n$, its radius of convergence is R. If $a_1 \in D(a, R)$, we have another power series

$$f_1(z) = \sum_{n=0}^{\infty} \frac{f^{(n)}(a_1)}{n!}(z - a_1)^n.$$

If its radius of convergence $R_1 \geq R - |a - a_1| > 0$, we call $f(z), f_1(z)$ as analytic elements, and $f_1(z)$ is an analytic continuation of $f(z)$. If we have m analytic elements $f_k(z) = \sum_{n=0}^{\infty} c_n^{(k)}(z - a_k)^n$ $(k = 1, 2, \cdots, m)$, f_k is the analytic continuation of f_{k-1}, then f_m is analytic continuation of f again. We start from $f(z)$, then we do it again and again, we obtain all analytic elements.

All these analytic elements forms a set. We call this set as a global analytic function.

A complete analytic (or complete holomorphic) function is a global analytic function which contains all analytic continuations of all its analytic elements. In general, it is a multivalued function. The union of circles of convergence of the analytic continuations is the existence domain of the complete analytic (or complete holomorphic) function. Of course, it cannot be analytically continued again. All the boundary points are singularities of the complete analytic (or complete holomorphic) function.

EXERCISES III

1. If $a_1, a_2, \cdots$ are distinct points, and $\lim_{n \to \infty} |a_n| = \infty$. If

$$\psi_n(z) = \sum_{j=1}^{\infty} \frac{c_{n,j}}{(z - a_n)^j}, \qquad n = 1, 2, \cdots,$$

are holomorphic on $\mathbb{C}$ except a_n $(n = 1, 2, \cdots)$, then there exists a holomorphic function $f(z)$ on $\mathbb{C} \backslash \{a_1, a_2, \cdots\}$ such that the principal part of the Laurent expansion of $f(z)$ at every a_n is $\psi_n(z)$ $(n = 1, 2, \cdots)$.

2. Prove Theorem 8 and Theorem 8'.

3. Expand each of the following function as Laurent series on its indicated domains.

(i) $\dfrac{1}{z^3(z + i)}, \quad 0 < |z + i| < 1;$

(ii) $\dfrac{z^2}{(z + 1)(z + 2)}, \quad 1 < |z| < 2;$

(iii) $\log \left(\dfrac{z - a}{z - b} \right), \quad \max(|a|, |b|) < |z| < +\infty;$

(iv) $z^2 e^{\frac{1}{z}}, \quad 0 < |z| < +\infty;$

(v) $\sin \dfrac{z}{1 + z}, \quad 0 < |z + 1| < +\infty.$

4. Find the singularities of each of the following functions. Point out the kinds of these singularities. Find its order if the singularity is a pole.

(i) $\dfrac{\sin z}{z};$ \qquad (ii) $\dfrac{1}{z^2 - 1} \cos \dfrac{\pi z}{z + 1};$ \qquad (iii) $z(e^{\frac{1}{z}} - 1);$

(iv) $\sin\dfrac{1}{1-z}$;   (v) $\dfrac{\exp\left(\dfrac{1}{1-z}\right)}{e^z-1}$;   (vi) $\tan z$.

5. Show that: (1) If a is an essential singularity of $f(z)$, and $f(z) \neq 0$, then a is an essential singularity of $\frac{1}{f(z)}$.

(2) If a is an essential singularity of $f(z)$, and $p(z)$ is a non-constant polynomial, then a is an essential singularity of $p(f(z))$.

6. Show that: (i) $\displaystyle\prod_{n=0}^{\infty}(1+z^{2^n}) = \frac{1}{1-z}$ when $|z| < 1$;

(ii) $\displaystyle \operatorname{sh}\pi z = \pi z \prod_{n=1}^{\infty}\left(1+\frac{z^2}{n^2}\right)$;

(iii) $\displaystyle \cos\pi z = \prod_{n=0}^{\infty}\left(1-\left(\frac{z}{n+1/2}\right)^2\right)$;

(iv) $\displaystyle e^z - 1 = ze^{\frac{z}{2}}\prod_{n=1}^{\infty}\left(1+\frac{z}{4\pi^2 n^2}\right)$.

7. (Blaschke product). Let the complex number sequence $\{a_k\}$ $(k = 1, 2, \cdots)$ satisfy $0 < |a_k| < 1$, and $\displaystyle\sum_{k=1}^{\infty}(1-|a_k|) < \infty$. Show that the infinite product

$$f(z) = \prod_{k=1}^{\infty}\frac{a_k - z}{1 - \overline{a}_k z}\cdot\frac{|a_n|}{a_n}$$

converges uniformly on $|z| \leq r$ $(0 < r < 1)$. Hence $f(z)$ is holomorphic on $|z| < 1$, a_k $(k = 1, 2, \cdots)$ are its zero points, $f(z)$ has no other zero point, and $|f(z)| \leq 1$.

8. (Poisson-Jensen formula) If $f(z)$ is a meromorphic function on $|z| \leq R$ $(0 < R < \infty)$, $a_1, a_2, \cdots, a_s$ and $b_1, b_2, \cdots, b_t$ are its zero points and poles on $|z| < R$ respectively. If a is a zero point of multiplicity n, then a appears at $a_1, a_2, \cdots, a_s, n$ times. Similarly if b is a pole of order m, then b appears at $b_1, b_2, \cdots, b_t, m$ times. For any point z in $|z| < R$, which is different from a_i $(i = 1, 2, \cdots, s)$ and b_j $(j = 1, 2, \cdots, t)$, show that the formula

$$\log|f(z)| = \frac{1}{2\pi}\int_0^{2\pi}\log|f(Re^{i\varphi})|Re\left(\frac{Re^{i\varphi}+z}{Re^{i\varphi}-z}\right)d\varphi$$

$$+ \sum_{j=1}^{t}\log\left|\frac{R^2 - \overline{b}_j z}{R(z - b_j)}\right| - \sum_{i=1}^{s}\log\left|\frac{R^2 - \overline{a}_j z}{R(z - a_j)}\right|$$

holds. This formula is the start point of Nevanlinna theory of value distribution.

9. If a meromorphic function $f(z)$ has two poles on the extended complex plane $\mathbb{C}^*$ only. One pole is $z = -1$ of order 1, its principal part is $\frac{1}{z+1}$, the other pole is $z = 2$ of order 2, its principal part is $\frac{2}{z-2} + \frac{3}{(z-2)^2}$, and $f(0) = \frac{7}{4}$. Find the Laurent expansion of $f(z)$ on $1 < |z| < 2$.

10. If a meromorphic function $f(z)$ has pole of order 2 at $z = 1, 2, 3, \cdots$, and the principal part of its Laurent expansion at the neighborhood of $z = n$ is $\frac{n}{(z-n)^2}$ $(n = 1, 2, \cdots)$. Find the general form of $f(z)$.

11. (i) Expand $f(z) = \dfrac{1}{e^z - 1}$ as a partial fraction;

(ii) Show that: $\dfrac{1}{\sin^2 \pi z} = \dfrac{1}{\pi^2} \displaystyle\sum_{n=-\infty}^{\infty} \dfrac{1}{(z-n)^2}$;

(iii) Show that: If $\alpha \neq 0$, $\dfrac{\beta}{\alpha} \neq \pm 1, \pm 2, \cdots$, then

$$\frac{\pi}{\alpha} \cot \frac{\pi \beta}{\alpha} = \sum_{n=0}^{\infty} \left\{ \frac{1}{n\alpha + \beta} - \frac{1}{n\alpha + (\alpha - \beta)} \right\},$$

and then to show that

$$\frac{1}{1 \cdot 2} + \frac{1}{4 \cdot 5} + \frac{1}{7 \cdot 8} + \cdots + \frac{1}{(3n-2) \cdot (3n-1)} + \cdots = \frac{\pi}{3\sqrt{3}}.$$

12. Suppose a meromorphic function $f(z)$ has a finite number of poles: $\alpha_1, \alpha_2, \cdots, \alpha_m$, and each $\alpha_k (1 \leq k \leq m)$ is not an integer, $z = \infty$ is a zero point of multiplicity p of $f(z)$. Show that:

(i) $\displaystyle\lim_{n \to \infty} \sum_{k=-n}^{n} f(k) = -\pi \sum_{k=1}^{m} \text{Res}\, (f(z)\cot \pi z, \alpha_k)$;

(ii) $\displaystyle\lim_{n \to \infty} \sum_{k=-n}^{n} (-1)^k f(k) = -\pi \sum_{k=1}^{m} \text{Res}\, \left(\frac{f(z)}{\sin \pi z}, \alpha_k \right)$;

(iii) Use (i) and (ii) to find the sum of each of the following series:

$$\sum_{n=-\infty}^{\infty} \frac{1}{(a+n)^2}, \quad \alpha \text{ is not an integer};$$

$$\sum_{n=0}^{\infty} \frac{(-1)^n}{n^2 + a^2}, \quad a \text{ is a non-zero real number}.$$

13. Find the residue of each of the following functions at its isolate singularities (including the point at infinity, if it is an isolate singularity).

(i) $\dfrac{1}{z^2 - z^4}$;   (ii) $\dfrac{z^2 + z + 2}{z(z^2 + 1)^2}$;

(iii) $\dfrac{z^{n-1}}{z^n + a^n}$, $a \neq 0$, n is a positive integer;

(iv) $\dfrac{1}{\sin z}$;   (v) $z^3 \cos \dfrac{1}{z-2}$;   (vi) $\dfrac{e^z}{z(z+1)}$.

14. Suppose functions $f(z), g(z)$ are holomorphic at $z = a$, $f(a) \neq 0$, $z = a$ is a zero point of multiplicity 2 of $g(z)$. Find $\operatorname{Res}\left(\dfrac{f(z)}{g(z)}, a\right)$.

15. Evaluate each of the following integrals

(i) $\displaystyle\int_0^{+\infty} \dfrac{x^2 dx}{(x^2 + 1)^2}$;   (ii) $\displaystyle\int_0^{\frac{\pi}{2}} \dfrac{dx}{a + \sin^2 x}$, $a > 0$;

(iii) $\displaystyle\int_0^{+\infty} \dfrac{x \sin x}{x^2 + 1} dx$;   (iv) $\displaystyle\int_0^{+\infty} \dfrac{\log x}{(x^2 + 1)^2} dx$;

(v) $\displaystyle\int_0^{+\infty} \dfrac{x^{1-\alpha}}{1 + x^2} dx$, $0 < \alpha < 2$;

(vi) $\displaystyle\int_0^{+\infty} \dfrac{dx}{1 + x^n}$, n is an integer greater than 1;

(vii) $\displaystyle\int_0^{\pi} \dfrac{d\theta}{a + \cos \theta}$, $a > 1$ is a constant;

(viii) $\displaystyle\int_0^{+\infty} \left(\dfrac{\sin x}{x}\right)^2 dx$;

(ix) $\displaystyle\int_0^{+\infty} \dfrac{x^p}{x^2 + 2x \cos \lambda + 1} dx$, $-1 < p < 1$, $-\pi < \lambda < \pi$;

(x) $\displaystyle\int_0^{+\infty} \dfrac{1}{1 + x^p} dx$, $p > 1$;

(xi) $\displaystyle\int_0^1 \dfrac{x^{1-p}(1 - x)^p}{1 + x^2} dx$, $-1 < p < 2$;

(xii) $\displaystyle\int_0^{+\infty} \dfrac{\log x}{x^2 + 2x + 2} dx$;

(xiii) $\displaystyle\int_0^{+\infty} \dfrac{\sqrt{x} \log x}{x^2 + 1} dx$;

(xiv) $\displaystyle\int_0^{+\infty} \log\left(\frac{e^x+1}{e^x-1}\right) dx;$

(xv) $\displaystyle\int_0^{+\infty} \frac{x}{e^x+1} dx;$

(xvi) $\displaystyle\int_0^{\frac{\pi}{2}} \log\sin\theta\, d\theta;$

(xvii) $\displaystyle\int_0^{\pi} \frac{x\sin x}{1-2a\cos x+a^2} dx, \quad a>0;$

(xviii) $\displaystyle\int_0^{+\infty} \frac{\log(1+x^2)}{1+x^2} dx.$

16. Can we analytically continued the function defined by the series $-\frac{1}{z}-1-z-z^2-\cdots$ on $0<|z|<1$ to the function defined by $\frac{1}{z^2}+\frac{1}{z^3}+\frac{1}{z^4}+\cdots$ on $|z|>1$? Explain the reason.

17. Show that the functions defined by the series

$$1+\alpha z+\alpha^2 z^2+\cdots+\alpha^n z^n+\cdots$$

and

$$\frac{1}{1-z}-\frac{(1-\alpha)z}{(1-z)^2}+\frac{(1-\alpha)^2 z^2}{(1-z)^3}$$

are analytic continuation of each other.

18. Show that the function $f_1(z)$ defined by the series

$$z-\frac{1}{2}z^2+\frac{1}{3}z^3+\cdots$$

on $|z|<1$, and the function $f_2(z)$ defined by the series

$$\ln 2-\frac{1-z}{2}-\frac{(1-z)^2}{2\cdot 2^2}-\frac{(1-z)^3}{3\cdot 2^3}-\cdots$$

on $|z-1|<2$, are analytic continuation of each other.

19. Show that the power series $\displaystyle\sum_{n=0}^{\infty} z^{2^n}$ can not be analytically continued to the outside of its circle of convergence.

CHAPTER IV
RIEMANN MAPPING THEOREM

§ 4.1 Conformal Mapping

Another important part of complex analysis is the theory of conformal mappings of Riemann. The key point of this theory is that we treat the holomorphic function $w = f(z)$ as a mapping from a domain in z-plane onto a domain in w-plane. That means, we treat the holomorphic function in the point veiw of geometry. In Chapter I §1.3, we mentioned that if $f'(z) \neq 0$, then the mapping $w = f(z)$ is conformal. So we call it as a conformal mapping or a holomorphic mapping.

First of all, we observe the following fact: If $U \subseteq \mathbb{C}$ is a domain, $w = f(z)$ is a holomorphic mapping, then $f(U)$ is a domain again.

We only need to show that $f(U)$ is a connected open set. If w_1, w_2 are any two points in $f(U)$, then we may find z_1, z_2 in U, so that $w_1 = f(z_1)$, $w_2 = f(z_2)$. Since U is connected, we have a curve $\gamma(t)$ in U which connect z_1 and z_2. Obviously $f(\gamma(t)) \subset f(U)$, and connects w_1 and w_2. Hence $f(U)$ is connected.

If w_0 is any point in $f(U)$, by Chapter II §2.4 Theorem 15, for sufficiently small $\rho > 0$, there exists a $\delta > 0$, so that for any point w in $D(w_0, \delta)$, we may find a point z in $D(z_0, \rho)$, such that $f(z) = w$. That means $D(w_0, \delta) \subset f(U)$. Hence $f(U)$ is an open set. This result is called as the open mapping theorem: f maps an open set onto an open set.

In Chapter I §1.5, we have already defined that: a function $f(z)$ is univalent on $U \subset \mathbb{C}$ if $f(z_1) = f(z_2)$ implies $z_1 = z_2$. We have the following result.

If $f(z)$ is univalent and holomorphic on $U \subset \mathbb{C}$, then $f'(z) \neq 0$ for any point $z \in U$. Conversely, if $f(z)$ is holomorphic on U, and $f'(z_0) \neq 0$ where $z_0 \in U$, then there exists a neighborhood of z_0, such that $f(z)$ is univalent on this neighborhood.

We prove this result as follows.

If $f(z)$ is univalent and holomorphic on $U \subseteq \mathbb{C}$, and there is a point

$z_0 \in U$, $f'(z_0) = 0$, then z_0 is the zero point of multiplicity m $(m \geq 2)$ of the function $f(z) - f(z_0)$. We may find a neighborhood $N(z_0)$ of z_0, so that $f'(z) \neq 0$ when $z \in N(z_0)$ except the point z_0. Let $N_1(w_0)$ be the corresponding neighborhood of $w_0 = f(z_0)$ in $f(U)$. For any $w \in N_1(w_0)$, the function $f(z) - w$ has exact m zero points in $N(z_0)$ by Chapter II §2.4 Theorem 15. It contradicts with the univalency of $f(z)$ on U. Conversely, if $f'(z_0) \neq 0$, then z_0 is a simple zero point of the function $f(z) - f(z_0)$. By Chapter II §2.4 Theorem 15, for any sufficiently small $\rho > 0$, there exists a $\delta > 0$, so that for any point $w \in D(f(z_0), \delta)$, $f(z) - w$ has one zero point only in $D(z_0, \rho)$. It means that there is unique z, $f(z) = w$. We choose $\rho_1 < \rho$, and $f(D(z_0, \rho_1)) \subset D(w_0, \delta)$. Hence $f(z)$ is univalent on $D(z_0, \rho)$.

Moreover, it is easy to prove that: If $w = f(z)$ is univalent and holomorphic on U, and maps U onto G, then the inverse function $z = g(w)$ is univalent and holomorphic on G, and maps G onto U. Thus, we call the univalent holomorphic mapping as the **biholomorphic mapping**.

Theorem 1   Suppose $G \subseteq \mathbb{C}$ is a domain, γ is a rectifiable simple closed curve in G, the inner points set bounded by γ is a domain $U \subset G$. If $f(z)$ is holomorphic on G, and it one to one maps γ to a simple closed curve Γ, then $w = f(z)$ is univalent on U, and it maps U onto the inner points set V bounded by Γ.

Proof   If w_0 is not on Γ, and the number of zero points of $f(z) - w_0$ in the inner points set bounded by γ is N, then N equals to

$$\frac{1}{2\pi i} \int_\gamma \frac{f'(z)}{f(z) - w_0} dz = \pm \frac{1}{2\pi i} \int_\Gamma \frac{dw}{w - w_0}$$

by Argument principle (Chapter II §2.4 Theorem 12). If w_0 is outside on the domain bounded by Γ, then

$$\int_\Gamma \frac{dw}{w - w_0} = 0.$$

Hence $N = 0$, and $f(z) - w_0$ has no zero point in U. If w_0 is in the domain bounded by Γ, then

$$\frac{1}{2\pi i} \int_\Gamma \frac{dw}{w - w_0} = 1.$$

Thus $\frac{1}{2\pi i} \int_\gamma \frac{f'(z)}{f(z) - w_0} dz = \pm 1$. Obviously N is non-negative, we have $N = 1$. Thus $f(z) - w_0$ has one zero point in U only. When z describes on γ in positive direction once, $w = f(z)$ describes on Γ in positive direction once also. When $w_0 \in \Gamma$, $f(z) - w_0$ has no zero point on U. If it is not true, there exists $z_0 \in U$, and $f(z_0) = w_0$, then we have $D(w_0, \delta) \subset f(U)$, and for every $w_1 \in D(w_0, \delta)$, $f(z) - w_1$ has a zero point in U. We may take $w_1 \in D(w_0, \delta)$ and w_1 is outside Γ. It is contradicted with no zero point of $f(z) - w_1$ in U.

The following are the simplest examples of conformal mappings.

Example 1   In Chapter II §2.5 we have proved that the univalent conformal mappings of unit disk $D(0, 1)$ onto itself are

$$w = e^{i\theta} \frac{z - a}{1 - \overline{a}z}, \qquad a \in D(0, 1), \quad \theta \in \mathbb{R}. \tag{1.1}$$

Besides these mappings, there are no other univalent conformal mappings with this property.

Example 2   The univalent conformal mappings of the upper half plane $\operatorname{Im} z > 0$ onto the unit disk are

$$w = e^{i\theta} \frac{z - a}{z - \overline{a}}, \qquad \operatorname{Im} a > 0, \quad \theta \in \mathbb{R}. \tag{1.2}$$

Besides these mappings, there are no other univalent conformal mappings with this property.

Obviously (1.2) maps $\operatorname{Im} z = 0$ onto $|w| = 1$. (1.2) can be expressed as

$$z = \frac{\overline{a}w - e^{i\theta}a}{w - e^{i\theta}}$$

which maps $|w| = 1$ onto $\operatorname{Im} z = 0$. From Theorem 1, this mapping maps $D(0, 1)$ onto $\operatorname{Im} z > 0$, and the mapping is univalent and holomorphic on $D(0, 1)$.

Conversely, if $w = f(z)$ is an univalent and holomorphic mapping of $\operatorname{Im} z > 0$ onto $D(0, 1)$, we already know that (1.2) maps $\operatorname{Im} z > 0$ onto $D(0, 1)$, and (1.2) is univalent and holomorphic on $\operatorname{Im} z > 0$. We denote (1.2) by $\psi(z)$. Then $f \circ \psi^{-1}$ maps unit disk to unit disk. By example 1, $f \circ \psi^{-1}$ is a function as (1.1), we denote it by φ, $f\psi^{-1} = \varphi$. Thus $f = \varphi \circ \psi$. It is a function as (1.2).

Similarly we may prove

Example 3   The univalent conformal mappings of the upper half plane

Im $z > 0$ onto upper half plane Im $w > 0$ are

$$w = \frac{az + b}{cz + d},$$

where a, b, c, d are real numbers, and $ad - bc > 0$. Besides these mappings, there are no other univalent conformal mappings with this property.

In Chapter III §3.3, we have proved that the set of all the linear fractional transformations $\{w = \frac{az+b}{cz+d}, \ a, b, c, d \in \mathbb{C}, \ ad - bc = 1\}$ forms the group of meromorphic automorphisms of the extended complex plane $\mathbb{C} \cup \{\infty\}$. We denote it by Aut $\{\mathbb{C}^*\}$. Moreover, we may establish the one to one correspondence between Aut $(\mathbb{C}^*)$ and the Möbius group $SL(2, \mathbb{C})/\{\pm I\}$, if we let $w = \frac{az+b}{cz+d}$ correspond with $\left(\begin{smallmatrix} a & b \\ c & d \end{smallmatrix}\right)/\{\pm I\}$.

Let $z = \frac{aw+b}{cw+d}$ be any linear fractional transformation. If we regard the straight line as a circle with radius ∞, then the linear fractional transformation has the following important property: The linear fractional transformation maps circle to circle. We prove this property as follows: Let $z = x + iy$, we may express any circle by

$$\alpha(x^2 + y^2) + \beta x + \gamma y + \delta = 0,$$

which $\alpha, \beta, \gamma, \delta$ are real numbers. We rewrite this equation as

$$\alpha z \bar{z} + \frac{1}{2}\beta(z + \bar{z}) + \frac{1}{2i}\gamma(z - \bar{z}) + \delta = 0.$$

It is

$$A z \bar{z} + B z + \overline{B} \bar{z} + C = 0, \qquad (1.3)$$

where $A = \alpha$, $C = \delta$ are real numbers, and $B = \frac{1}{2}\beta + \frac{1}{2i}\gamma$ is a complex number. If $\alpha = 0$, i.e. $A = 0$, then (1.3) is a straight line, otherwise (1.3) is a circle. Any linear fractional transformation $z = \frac{aw+b}{cw+d}$ is composed by the translation $z = w + b$, the dialation $z = aw$ and the invertion $z = \frac{1}{w}$. It is easy to verity that any one of those three transformations maps a circle to a circle.

If we substitute $z = w + b$ into (1.3), we have

$$Aw\bar{w} + (A\bar{b} + B)w + (Ab + \overline{B})\bar{w} + Ab\bar{b} + Bb + \overline{B}\bar{b} + C = 0.$$

It is a mapping of (1.3).

If we substitute $z = aw$ into (1.3), we have

$$Aa\bar{a}w\bar{w} + Baw + \overline{Ba}\,\bar{w} + C = 0.$$

It is a mapping of (1.3).

Finally, if we substitute $z = \frac{1}{w}$ into (1.3), we have

$$Cw\bar{w} + \overline{B}w + B\bar{w} + A = 0.$$

It is a mapping of (1.3). Thus the linear fractional transformation maps a circle to a circle.

If z_1, z_2, z_3, z_4 are four points in $\mathbb{C}^*$, at least three points of them are different. We call

$$(z_1, z_2, z_3, z_4) = \frac{(z_1 - z_3)(z_2 - z_4)}{(z_1 - z_4)(z_2 - z_3)}$$

as the **cross ratio** of these four points. If any one of these four points is ∞, we may define the cross ratio by limit. For example

$$(\infty, z_2, z_3, z_4) = \frac{z_2 - z_4}{z_2 - z_3}.$$

We may prove that: if the linear fractional transformation $w = \frac{az+b}{cz+d}$ maps z_1, z_2, z_3, z_4 to w_1, w_2, w_3, w_4 respectively, then

$$(w_1, w_2, w_3, w_4) = (z_1, z_2, z_3, z_4).$$

That means: the cross ratio is invariant under linear fractional transformations. To prove it is easy, we only need to substitute $w = \frac{az+b}{cz+d}$ dirctly into (w_1, w_2, w_3, w_4), after simple computation, we get (z_1, z_2, z_3, z_4). Or, we may prove it by the following way. Let $w_i = \frac{az_i+b}{cz_i+d}$ $(i = 2, 3, 4)$. Then $(z, z_2, z_3, z_4) = (w, w_2, w_3, w_4)$ is a Möbius transformation, which maps z_2 to w_2, z_3 to w_3 and z_4 to w_4 respectively. Any Möbius transformation is decided by three points. Hence it is $w = \frac{az+b}{cz+d}$. Thus $w_1 = \frac{az_1+b}{cz_1+d}$, and $(w_1, w_2, w_3, w_4) = (z_1, z_2, z_3, z_4)$.

Conversely, if we have a function $f(z_1, z_2, z_3, z_4)$ which is an invariant under the group of linear fractional then f is a function of cross ratio. Under this understanding, the only invariant under the group of linear fractional transformations is the cross ratio. We may prove it as follows. Let T denote a linear fractional transformation. By assumption,

$$f(Tz_1, Tz_2, Tz_3, Tz_4) = f(z_1, z_2, z_3, z_4)$$

holds for any T. Let $Tz = (z, z_2, z_3, z_4)$, then

$$f(z_1, z_2, z_3, z_4) = f\left(\frac{(z_1 - z_3)(z_2 - z_4)}{(z_1 - z_4)(z_2 - z_3)}, 1, 0, \infty\right) = f((z_1, z_2, z_3, z_4), 1, 0, \infty).$$

We have proved the result.

§ 4.2 Normal Family

In the Riemann conformal mapping theory, the most important and deepest theorem is the Riemann mapping theorem

Theorem 2 (Riemann mapping theorem) Let $U \subseteq \mathbb{C}$ be a simply connected domain. If its boundary point is more than one, and z_0 is any point in U, then there exists an unique univalent holomorphic function $f(z)$ on U onto $D(0, 1)$ with $f(z_0) = 0$, $f'(z_0) > 0$.

The requirement about the boundary points in the theorem is natural. If the domain has one boundary point only, without loss of generality, we may assume it is point at infinity. If $f(z)$ maps it to the unit disk, then it is a constant by Liouville theorem (Chapter II, §2.3, Theorem 8).

From this theorem, we immediately have the following consequence: Any two simply connected domains in $\mathbb{C}$, if their boundary points are more than one, then they can univalent holomorphic maps to each other.

Let U and V be any two domains in $\mathbb{C}$. If there exists an univalent holomorphic function mapping U onto V, then we call U and V are **holomorphic equivalent**. The Riemann mapping theorem tells us: All simply connected domains with more than one boundary point are holomorphic equivalent to each other. Obviously, all simply connected domains are topologic equavalent, that is, there are continuous mappings which maps one domain to another. The Riemann mapping theorem tells us: topological equivalent implies holomorphic equivalent. Thus the Riemann mapping theorem is a very deep theorem. In Chapter VI, we will show that, this theorem is never hold in the high dimensional case (Chapter VI §6.4, Theorem 11, Poincaré theorem). Hence the Riemann mapping Theorem has a very special position in the theory of function of one complex variable.

The proof of this theorem depends on the idea of normal family. The idea of normal family is a basic idea in the theory of functions. In some sense, it corresponds the compact set in set theory. We will discuss it again in the Chapter V.

Definition 1   A family $\mathcal{F}$ of functions on U is normal, if any sequence in $\mathcal{F}$, we may select an uniformly convergence subsequence on any compact subset in U.

Theorem 3 (Montel theorem)   Let $U \subseteq \mathbb{C}$ be a domain and $\mathcal{F}$ be a family of holomorphic functions on U. If there exists a positive constant M, such that

$$|f(z)| \leq M$$

holds for any $z \in U$, $f \in \mathcal{F}$, then $\mathcal{F}$ is a normal family.

In purpose to prove Theorem 3, we need to prove the following Theorem 4. It is the well-known Ascoli-Arzela Theorem. It is a very useful theorem. We will use it to prove Theorem 5 of Chapter V §5.5.

Definition 2   Let $\mathcal{F} = \{f\}$ be a family of functions on the domain $S \subseteq \mathbb{R}^n$. If for any $\varepsilon > 0$, we may find a $\delta > 0$, such that

$$|f(z) - f(w)| < \varepsilon$$

holds for any $f \in \mathcal{F}$, and any two points $z, w \in S$ with $|z - w| < \delta$, then $\mathcal{F}$ is equicontinuous.

Definition 3   Let $\mathcal{F} = \{f\}$ be a family of functions on the domain $S \subseteq \mathbb{R}^n$. If there exists a positive number $M > 0$, such that

$$|f(z)| \leq M$$

holds for any $z \in S$, $f \in \mathcal{F}$, then $\mathcal{F}$ is equibounded.

Theorem 4 (Ascoli-Arzela theorem)   Let K be a compact set in $\mathbb{R}^n$. If the family of functions $\mathcal{F} = \{f_\nu\}$ is equicontinuous and equibounded, then there exists a subsequence in $\mathcal{F}$, which converges uniformaly on K.

That means: on compact set, equicontinuous and equibounded imply uniformly convergence.

Proof   There exists an everywhere densed sequence $\{\zeta_k\}$ on K, for example, all the points with rational coordinates. Since $\mathcal{F}$ is equibounded on K, for ζ_1, we may find a convergence subsequence $\{f_{\nu_{1k}}(\zeta_1)\}$ from sequence $\{f_\nu(\zeta_1)\}$. Then we may find a convergence subsequence from $\{f_{\nu_{1k}}(\zeta_2)\}$, and denote it by $\{f_{\nu_{2k}}(\zeta_2)\}$, $\cdots$, and so on. We have an array

$$
\begin{aligned}
&\nu_{11} < \nu_{12} < \cdots < \nu_{1j} < \cdots, \\
&\nu_{21} < \nu_{22} < \cdots < \nu_{2j} < \cdots, \\
&\quad\cdots \\
&\nu_{k1} < \nu_{k2} < \cdots < \nu_{kj} < \cdots, \\
&\quad\cdots.
\end{aligned} \tag{2.1}
$$

Each row is a subsequence of the previous one, and $\lim\limits_{j\to\infty} f_{\nu_{kj}}(\zeta_k)$ exist for all k. Obviously ν_{jj} is a strictly increasing sequence, and $\{\nu_{jj}\}$ is a subsequence of each row. Hence $\{f_{\nu_{jj}}\}$ is a subsequence of $\{f_\nu\}$, and it converges at all ζ_k, $k = 1, 2, \cdots$. For simplify, we denote ν_{jj} by ν_j.

We assume that $\mathcal{F}$ is equicontinuous on K. For any $\varepsilon > 0$, we may find a $\delta > 0$, so that for any two points $z, z' \in K$, and $f \in \mathcal{F}$, we have $|f(z) - f(z')| < \varepsilon/3$ if $|z - z'| < \delta$. Since K is a compact set, we may find a finite number of neighborhoods with radius $\delta/2$, which cover K. Selecting one point ζ_k in each neighborhood, there exist a positive integer N, such that

$$
|f_{\nu_i}(\zeta_k) - f_{\nu_j}(\zeta_k)| < \frac{\varepsilon}{3}
$$

holds when $i, j > N$. For any point $z \in K$, we may find a point ζ_k with $|\zeta_k - z| < \delta$, hence

$$
|f_{\nu_i}(z) - f_{\nu_i}(\zeta_k)| < \frac{\varepsilon}{3}
$$

and

$$
|f_{\nu_j}(z) - f_{\nu_j}(\zeta_k)| < \frac{\varepsilon}{3}
$$

hold. Thus, we have

$$
|f_{\nu_i}(z) - f_{\nu_j}(z)| < \varepsilon
$$

when $i, j > N$. We conclude that $\{f_{\nu_j}\}$ converges uniformly on K since K is a compact set.

We have proved the theorem.

We make two remarks on this theorem.

(1) The conditions in Theorem 4, equicontinuous and equibounded are the sufficient conditions for uniformly convergence, but it is the necessary condition for uniformly convergence also. We omit the detail of the proof.

(2) We used the Euclidean metric in the definition of equicontinuous and Theorem 4. If we replace Euclid metric by other metric (cf. Chapter V §5.1) in the definition 2, then Theorem 4 holds true again. We will use this result in Theorem 7 of Chapter V §5.5. We omit the detail of the proof.

Now we use Ascoli-Arzela theorem (Theorem 4) to prove Montel theorem (Theorem 3).

Proof of Theorem 3   Fix a point $z_0 \in U$, we may choose $R > 0$, so that $\overline{D}(z_0, R) \subseteq U$. The complement U^c of U is a closed set since U is open. $\overline{D}(z_0, R)$ and U^c are non-intersected. There is a positive distance between these two closed sets. There exists a $c > 0$, such that $|z - u| > c$ for any $z \in \overline{D}(z_0, R)$, $u \in U^c$. For any $z \in \overline{D}(z_0, R)$, and any $f \in \mathcal{F}$, $|f'(z)| \leq \frac{M}{c}$ by using Cauchy inequality on $\overline{D}(z, c)$. Denote $\frac{M}{c} = C$, for any two points $z, w \in \overline{D}(z_0, R)$, we have

$$|f(z) - f(w)| \leq C(z - w|.$$

It means that $\mathcal{F}$ is equicontinuous on $\overline{D}(z_0, R)$. In fact, for any $\varepsilon > 0$, we only need to take $\delta = \varepsilon/C$.

If K is any compact subset in U. We may find a finite number of $\overline{D}(z_0, R)$ to cover K. Hence $\mathcal{F}$ is equicontinuous on K. By Ascoli-Arzela theorem (Theorem 4), for any sequence $\{f_\nu\}$ in $\mathcal{F}$, we may find a subsequence $\{f_{\nu_k}\}$ which converges uniformly on K. We may use the diagonal method which we used in the proof of Theorem 4, to prove that there exists a subsequence $\{f_{\nu_k}\}$ in the sequence $\{f_\nu\}$, which converges uniformly on any compact subset in U.

We have proved the theorem.

In Chapter V §5.5, we will extend the idea of normal family and Montel theorem, and use it to prove the well-known Picard theorem.

§ 4.3   Riemann Mapping Theorem

Now we use Montel theorem (Theorem 3) to prove the Riemann mapping theorem (Theorem 2).

Proof of Theorem 2   We assume U is a bounded domain at first. Fix a point $z_0 \in U$, and denote the family of univaient holomorphic function by $\mathcal{F} = \{\sigma(z)\}$, where $\sigma(z)$ is an univalent holomorphic function of U into $D(0, 1)$ with $\sigma(z_0) = 0$. We observe that $\mathcal{F}$ is non-empty. Since U is bounded, we may find $R > 0$ so that $U \subseteq D(0, R)$, the function $\sigma(\zeta) = \frac{1}{2R}(\zeta - z_0)$ maps z_0 to 0, holomorphic and univalent, and satisfies $|\sigma(\zeta)| < \frac{1}{2R}(R+R) = 1$ on U. Hence $\sigma \in \mathcal{F}$, and $\mathcal{F}$ is non-empty. Every function in $\mathcal{F}$ is holomorphic and bounded (upper bound is 1). It implies that $\mathcal{F}$ is a normal family by Montel theorem (Theorem 3). Let

$$M = \sup\{|\sigma'(z_0)| \mid \sigma \in \mathcal{F}\}.$$

If $\overline{D}(z_0, r)$ is a closed disk centered at z_0 with radius r in U, we have $M \le \frac{1}{r}$ since $|\sigma'(z_0)| \le \frac{1}{r}$ by Cauchy inequality. We need to prove that there exists a σ_0 in $\mathcal{F}$, so that $\sigma_0'(z_0) = M$.

By the definition of M, there exists a sequence $\{\sigma_j\}$ in $\mathcal{F}$, $|\sigma_j'(z_0)| \to M$ when $j \to \infty$. Since $\mathcal{F}$ is a normal family, there exists a subsequence $\{\sigma_{j_k}\}$ in $\{\sigma_j\}$, and $\{\sigma_{j_k}\}$ converges uniformly to σ_0 on any compact subset in U. Since $|\sigma_{j_k}'(z_0)| \to M$ when $j_k \to \infty$, we have $|\sigma_0'(z_0)| = M$. Multiplying a complex number with modulus one to σ_0, we have a new σ_0 which has the equality $\sigma_0'(z_0) = M$.

Now we try to prove that σ_0 is univalent on U. Here we need to use the priniciple of argument (Chapter II §2.4, Theorem 12). If P, Q are any two different points in U, and $0 < s < |P - Q|$. Consider the function $\psi_k(z) \equiv \sigma_{j_k}(z) - \sigma_{j_k}(Q)$ on $\overline{D}(P, s)$. σ_j is univalent, and ψ_k is non-zero on $\overline{D}(P, s)$. The limiting function of $\psi_k(z)$ is $\sigma_0(z) - \sigma_0(Q)$ which is identically zero or identically non-zero by Hurwitz Theorem (Chapter II §2.4, Theorem 13). σ_0 is impossible identically zero due to $\sigma_0'(z_0) = M > 0$. Hence, for any $z \in \overline{D}(P, s)$, we have $\sigma_0(z) \ne \sigma_0(Q)$. In particular $\sigma_0(P) \ne \sigma_0(Q)$. P, Q are two arbitrary points in U, we have proved the univalency of σ_0.

Finally, we will prove that σ_0 maps U onto $D(0, 1)$.

If F is a non-zero holomorphic function on U, we may define $\log F$ on U by

$$\log F(z) = \int_{\gamma_z} \frac{F'(\zeta)}{F(\zeta)} d\zeta + \log F(z_0),$$

where γ_z is a piecewise C^1 curve in U from z_0 to z. The definition of $\log F(z)$ is independent of the choice of γ_z since U is simply connected. We may define the α-th power of $F(z)$ by $\log F(z)$ as

$$F^\alpha(z) = \exp(\alpha \log F(z)),$$

where $\alpha \in \mathbb{C}$.

If σ_0 does not map U onto $D(0,1)$, there exists a point $\beta \in D(0,1)$, and $\beta \notin \sigma_0(U)$. Let

$$\varphi_\beta(\zeta) = \frac{\zeta - \beta}{1 - \overline{\beta}\zeta},$$

then $\mu(\zeta) = (\varphi_\beta \circ \sigma_0(\zeta))^{\frac{1}{2}}$ is a holomorphic function on U. Let $\tau = \mu(z_0)$,

$$\varphi_\tau(\zeta) = \frac{\zeta - \tau}{1 - \overline{\tau}\zeta}, \qquad \eta(\zeta) = \varphi_\tau \circ \mu(\zeta)$$

and

$$\nu(\zeta) = \frac{|\eta'(z_0)|}{\eta'(z_0)}\,\eta(\zeta),$$

then $\nu \in \mathcal{F}$, $\nu(z_0) = 0$ and

$$|\nu'(z_0)| = \frac{1 + |\beta|}{2|\beta|^{\frac{1}{2}}}\,M > M$$

which is contradicted with the definition of σ_0. Thus σ_0 is an onto mapping. The function σ_0 is the function which we need in the Riemann mapping theorem. If there is another function g with the same properties, then $G(z) = f(g^{-1}(z))$ is an automorphism of the unit disk, and $G(0) = 0$, $G'(0) > 0$. We have $G(z) = z$ by Theorem 18 of Chapter II §2.5.

If U is an unbounded domain, we may transform U to a bounded domain.

We assume that the boundary points of U have two points at least, and 0, a $(\neq \infty)$ are two boundary points, otherwise we may use a linear fractional transformation to transform two boundary points to 0 and a.

Taking one single-valued branch of $\sqrt{z-a}$ on U, we denote it by $g(z)$, $g(U)$ is simply connected since U is simply connected. Of course, $g(z)$ is univalent on U. If it is not true, there exist $z_1, z_2 \in U$ such that $z_1 \neq z_2$ and $\sqrt{z_1 - a} = \sqrt{z_2 - a}$, then $z_1 - a = z_2 - a$, and hence $z_1 = z_2$, we get a contradiction. Next we prove $g(U) \cap (-g(U)) = \emptyset$. If it is not true, there exists a point $P \in g(U)$, and $-P \in g(U)$, then there exist $z_1, z_2 \in D$, so that $\sqrt{z_1 - a} = P$, $\sqrt{z_2 - a} = -P$. It implies $z_1 = z_2$, and hence $P = -P$, $P = 0$. But $0 \in \partial U$. We get a contradiction.

For any point $q \in g(U)$, there exists a neighboorhood $U_q \subset g(U)$ of point $g(q)$, and $-U_q \not\subset g(U)$ since $g(U)$ is simply connected. Taking a point b in $-U_q$, and let $\varphi(z) = \frac{1}{z-b}$, then $\varphi(z)$ maps $g(U)$ onto a bounded simply connected domain. Thus $\varphi \circ g$ is a univalent mapping which maps U onto a bounded simply connected domain.

The theorem have proved.

Riemann mapping theorem tells us that there exists a univalent holomorphic function which establishes the one-to-one correspondence between the points in the simply connected domain U and the points in the unit disk. Does there exist some correspondence between their boundary points? The boundary of a simply connected domain may be very complicated. Here we state a result about the simplest case and omit the detail of the proof. If U is a domain which is bounded by a Jordan curve Γ, and $w = f(z)$ is a univalent holomorphic function of U onto unit disk $D(0,1)$, then we may extend $f(z)$ to Γ, such that $f(z)$ is continuous on $\overline{U}$, and it astablishes the one-to-one correspondence between the points on Γ and the points on the unit circle $|w| = 1$.

§ 4.4 Symmetric Principle

Theorem 5 (Painlevé theorem) Let U_1, U_2 be two domains, $U_1 \cap U_2 = \emptyset$, $\partial U_1 \cap \partial U_2 = \gamma_0$ (Figure 8), where γ_0 is a rectifiable arc. If f_1, f_2 are holomorphic on U_1, U_2 respectively, continuous on $U_1 \cup \gamma$, $U_2 \cup \gamma$ respectively, and $f_1(z) = f_2(z)$ on γ where γ is γ_0 excluding its end points, then the function

$$f(z) = \begin{cases} f_1(z), & \text{when } z \in U_1, \\ f_1(z) = f_2(z), & \text{when } z \in \gamma, \\ f_2(z), & \text{when } z \in U_2 \end{cases}$$

is holomorphic on $U_1 \cup U_2 \cup \gamma$, f_1, f_2 are called mutual analytic continuation to each other across the boundary γ.

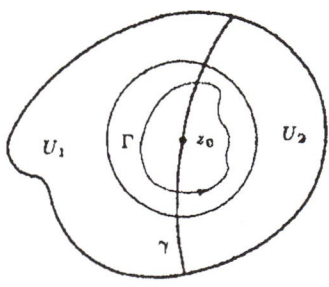

Fig. 8

Proof   We need only to show that f is holomorphic on γ

Suppose $z_0 \in \gamma$, we may select a r so that $D(z_0, r) \subset U = U_1 \cup U_2 \cup \gamma$. Let Γ be any rectifiable simple closed curve in $D(z_0, r)$. If Γ is inside $U_1 \cup \gamma$, then

$$\int_\Gamma f(z)\,dz = \int_\Gamma f_1(z)\,dz = 0$$

by Cauchy integral Theorem. Similarly, if Γ is inside $U_2 \cup \gamma$, then

$$\int_\Gamma f(z)\,dz = \int_\Gamma f_2(z)\,dz = 0.$$

If Γ lies on U_1 and U_2 both, Γ_1 is the part of Γ inside U_1, Γ_2 is the part of Γ inside U_2, Γ_0 is the part of γ inside the domain which is bouned by Γ, then

$$\int_\Gamma f(z)\,dz = \int_{\Gamma_1 + \Gamma_0} f_1(z)\,dz + \int_{\Gamma_2 - \Gamma_0} f_2(z)\,dz = 0.$$

Thus $f(z)$ is holomorphic on $D(z_0, r)$ by Morera theorem (Chapter II §2.3 Theorem 7). In particular, $f(z)$ is holomorphic at $z = z_0$. Since z_0 is an arbitrary point on γ, we have proved that $f(z)$ is holomorphic on U.

From Painlevé theorem, we have the following theorem.

Theorem 6 (Symmetric principle)   Suppose U is a domain, it locates at one side of the real axis, and its boundary contains a line segment s_0 on the real axis. If $f(z)$ is holomorphic on U, continuous on $U \cup s_0$, and $f(z)$ take real values on s_0, then there exists a function $F(z)$, which is holomorphic on $U \cup U' \cup s$, and $F(z) = f(z)$ on U, where U' is a symmetric domain of U with respect to the real axis, s is s_0 excluding its end points. Moreover, $F(\bar{z}) = \overline{F(z)}$.

Proof   We define a function $F(z)$ on $U \cup U' \cup s$ as

$$F(z) = \begin{cases} f(z), & \text{when } z \in U \cup s, \\ \overline{f(\bar{z})}, & \text{when } z \in U'. \end{cases}$$

The function satisfies $F(\bar{z}) = \overline{F(z)}$. We need to prove that $F(z)$ is holomorphic on $U \cup U' \cup s$. We prove $F(z)$ is holomorphic on U' at first. If $z_0 \in U'$, z is a point in the neighborhood of z_0, then

$$\frac{F(z) - F(z_0)}{z - z_0} = \frac{\overline{f(\bar{z})} - \overline{f(\bar{z}_0)}}{z - z_0} = \overline{\left(\frac{f(\bar{z}) - f(\bar{z}_0)}{\bar{z} - \bar{z}_0} \right)}.$$

Thus

$$\lim_{z \to z_0} \frac{F(z) - F(z_0)}{z - z_0} = \overline{f'(\overline{z}_0)}.$$

Since $f(z)$ has real values on s, $\overline{f(x_0)} = f(x_0)$ when $x_0 \in s$, we have

$$\lim_{\substack{x \to x_0 \\ x \in U'}} F(x) = \lim_{\substack{x \to x_0 \\ x \in U'}} \overline{f(\overline{z})} = \overline{\lim_{\substack{x \to x_0 \\ x \in U'}} f(\overline{z})} = \overline{f(x_0)} = f(x_0).$$

Hence $F(z)$ is continuous on $U' \cup s$. Thus $F(z)$ is holomorphic on $U \cup U' \cup s$ by Painlevé theorem (Theorem 5). We may state theorem 6 in the more general form.

Theorem 6′ (Symmetric principle)   Suppose U is a domain, it locates at one side of the straight line l, and its boundary contains a line segment s_0 on l. If $f(z)$ is holomorphic on U, continuous on $U \cup s_0$, and $f(z)$ takes values on a straight line L when $z \in s_0$, then there exists a function $F(z)$, which is holomorphic on $U \cup U' \cup s$, and $F(z) = f(z)$ on U, where U' is a symmetric domain of U with respect to l, s is s_0 excluding its end points. Moreover, if z_1, z_2 are two points in $U \cup U' \cup s$, which are symmetric with respect to l, then $F(z_1), F(z_2)$ are symmetric points with respect to L.

Proof   Using transformation $Z = az + b$ transforms l to the real axis, and using $W = cw + b$ transforms L to the real axis, we apply the symmetric principle to Z and W, after that we pull back to z and w.

We may extend the symmetric principle as follows. If we replace the line segments l and L in Theorem 6′ by circle arcs, then we may use the analytic continuation process again.

§ 4.5   Examples of Riemann Surface

Riemann mapping theorem tells us that any two simply connected domains (more than one boundary point) are holomorphic equivalent to each other. That means: there exists a one to one single valued (i.e. univalent) holomorphic mapping, which maps one to another one. If the holomorphic mapping is not one-to-one single valued (i.e. not univalent), how to establish the one-to-one correspondence between the domain and its image. This is the idea of Riemann surface. The idea of Riemann surface is one of the most important and basic ideas in complex analysis. There ara many very nice books about this topic. For example, L.V. Ahlfors and L. Sario[2], H. Wu, I.N. Lu and Z.H. Chen[3], G. Springer[4]. As a matter of fact, the Riemann surface is a complex manifold of one dimension. In general, complex analysis means to discuss analysis on complex manifold. As a textbook of complex analysis for undergraduate studtents, we simply describe what is the Riemann surface by example only.

When we discuss the elementary functions in Chapter I §1.5, we already know that, for the power function $w = z^\alpha$, $\alpha = a + \mathrm{i}\,b$, $w = z^\alpha$ is an infinite-valued function if $b \neq 0$; $w = z^\alpha$ is a single-valued function if $b = 0$ and a is an integer n, but its inverse function is not single-valued. The function $w = z^n$ maps an anguler domain $\frac{(k-1)2\pi}{n} < \arg z < \frac{k2\pi}{n}$ ($k = 1, 2, \cdots, n$), onto whole w-plane with a slit, and the slit is the positive real axis. The mapping is one-to-one and holomorphic. The image of each angular domain is w-plane with a "cut". Then these n angular domains are corresponded to the n w-complex planes with cut. Now we arrange these n w-complex planes altogether by the following way: identifing the lower side of the cut of the previous complex plane to the upper side of the cut of the next complex plane, the lower side of the cut of the nth complex plane identify to the upper side of the cut of the first complex plane. It forms a Riemann surface. We call each complex plane as one sheet of the Riemann surface, or the branch of the Riemann surface. It is clear that when z varifies at the z-complex plane, then w varifies at the Riemann surface. It establishs a one to one correspondence between the z-complex plane and the Riemann surface.

We may replace the cut, the positive real axis, by any ray from 0 to ∞ to get a new Riemann surface. It is identical with the original Riemann surface. We need to declare what kind cut we take when we consider Riemann surface.

The point $w = 0$ has a special meaning. It connects with all branchs. A closed curve around $w = 0$ need rotate n times. This point is called as **branch point**. If we consider the point at infinity, then point at infinity is a branch point also. In general, a branch point need not connected with all branchs. If it is connected with h sheets, then we call it as a branch point of order $h - 1$.

Similarly, we may consider the Riemann surface of $w = e^z$. The function maps a strip $(k - 1)2\pi < y < k2\pi$ ($z = x + \mathrm{i}\,y$, $k = 1, 2, \cdots$) to a sheet on the w complex plane with the positive real axis as it cut. There are infinite many sheets of the Riemann surface. The point $w = 0$ is not on the Riemann surface since e^z is never equal to zero.

Conversely, if n is a positive integer greater than 1, the function $w = z^{\frac{1}{n}}$

maps a n sheets Riemann surface to the w complex plane, and it establishes a one to one correspondence between them. Similarly, the function $w = \log z$ maps a Riemann surface with infinite many sheets to the w complex plane, and it establishes a one to one correspondence between them.

Thus, if we discuss the correspondence between the Riemann surface and the complex plane, we should indicate what sheet of the Riemann surface is discussed.

§ 4.6   Schwarz-Christoffel Formula

The Riemann mapping theorem is an existence theorem. How to give the mapping explicitly is not so easy. In §4.1, we gave some simplest examples. Now we will give the concrete formula of the mapping which maps the upper half plane to a polygon. This is the Schwarz-Christoffel formula.

Let $a_1, a_2, \cdots, a_n$ be n real numbers with $-\infty < a_1 < a_2 < \cdots < a_n < \infty$. We may denote $a_0 = -\infty$, and $a_{n+1} = +\infty$. If $\alpha_1, \alpha_2, \cdots, \alpha_n$ are n positive real numbers satisfing the condition

$$\alpha_1 + \alpha_2 + \cdots + \alpha_n + 1 < n.$$

Let

$$\beta(t) = (t - a_1)^{\alpha_1 - 1} \cdots (t - a_n)^{\alpha_n - 1}, \tag{6.1}$$

then

$$\int_{-\infty}^{\infty} |\beta(t)| \, dt < \infty.$$

When $t < a_k$, we may take a branch of

$$(t - a_k)^{\alpha_k - 1} = \exp\left((\alpha_k - 1) \log (t - a_k)\right)$$

with its argument $\pi(\alpha_k - 1)$. Thus

$$\arg \beta(t) = \pi\big[(\alpha_1 + \cdots + \alpha_n) - n\big]$$

when $t < a_1$;

$$\arg \beta(t) = \pi\big[(\alpha_k + \cdots + \alpha_n) - (n - k + 1)\big], \qquad 2 \le k \le n$$

when $t \in (a_{k-1}, a_k)$;

$$\arg \beta(t) = 0$$

when $t \in (a_n, \infty)$.

We define $n + 2$ complex numbers as

$$w_k = c \int_0^{a_k} (t - a_1)^{\alpha_1 - 1} \cdots (t - a_n)^{\alpha_n - 1} \, dt, \qquad 0 \le k \le n + 1 \qquad (6.2)$$

where c is a positive constant. In the upper half plane $\overline{H} = \{z \in \mathbb{C} \mid \operatorname{Im} z \ge 0\}$, we define the function

$$f(z) = c \int_0^z \beta(t) \, dt. \qquad (6.3)$$

It is a holomorphic function on $H = \{z \in \mathbb{C} \mid \operatorname{Im} z > 0\}$, and on the real axis,

$$f(x) = w_{k-1} + c \int_{a_{k-1}}^x \beta(t) \, dt$$

$$= w_{k-1} + c \, e^{i\,[(\alpha_k - 1)\pi + \cdots + (\alpha_n - 1)\pi]} \int_{a_{k-1}}^x |\beta(t)| \, dt$$

when $x \in (a_{k-1}, a_k)$ $(1 \le k \le n + 1)$. Then the function $f(x) - w_{k-1}$ has the same argument $[(\alpha_k - 1)\pi + \cdots + (\alpha_n - 1)\pi]$ on the interval (a_{k-1}, a_k), and its modulus is increasing from 0 to

$$l_k = c \int_{a_{k-1}}^{a_k} |\beta(t)| \, dt. \qquad (6.4)$$

Thus when x is varified in the interval $[a_{k-1}, a_k]$, f is varified in the interval $\Delta_{k-1} = [w_{k-1}, w_k]$, the argument of Δ_{k-1} is $(\alpha_k - 1)\pi + \cdots + (\alpha_n - 1)\pi$, and its length is l_k.

Now we prove that $w_0 = w_{n+1}$. We only need to prove that, for any $\varepsilon > 0$, there exists a $R > 0$, so that $|w_0 - f(z)| \le \varepsilon$ holds when $z \in \overline{H}$, $|z| \ge R$. If it is true, then it implies $\lim_{z \to \infty} f(z) = w_{n+1} = w_0$. We prove it as follows.

We know that

$$\int_{-\infty}^{\infty} |\beta(t)| \, dt < \infty.$$

For any $\varepsilon > 0$, there exists a $R_1 > 0$ such that

$$\int_{-\infty}^{-R_1} |\beta(t)| \, dt \le \frac{\varepsilon}{2}.$$

We have

$$\left| w_0 - c \int_0^x \beta(t) \, dt \right| \le \int_{-\infty}^{-R_1} |\beta(t)| \, dt \le \frac{\varepsilon}{2}$$

when $-\infty < x < -R_1$. Of course, we may chooce $R_1 \geq \max\{|a_1|, \cdots, |a_n|\}$. Let $z_0 = \rho_0 e^{i\theta_0}$, $\rho_0 \geq R_1$, $0 \leq \theta_0 \leq \pi$, then

$$|f(z_0) - f(\rho_0)| = c\left|\int_0^{\theta_0} (\rho_0 e^{i\theta} - a_1)^{\alpha_1-1} \cdots (\rho_0 e^{i\theta} - a_n)^{\alpha_n-1} \rho_0\, e^{i\theta}\, d\theta\right|$$

$$\leq c\rho_0(\rho_0 - R_1)^{\alpha_1+\cdots+\alpha_n-n}.$$

The right-hand side of the previous inequality approaches to zero when $\rho_0 \to \infty$ since $\alpha_1+\cdots+\alpha_n < n-1$. Thus there exists a $R \geq R_1$ so that $|f(z_0)-f(\rho_0)| \leq \varepsilon/2$ when $\rho_0 = |z_0| \geq R$. Hence $|f(z) - w_0| \leq \varepsilon$ when $|z| \geq R$, $\mathrm{Im}\, z \geq 0$. We have proved that $w_0 = w_{n+1}$. We conclude that f maps $\mathbb{R} \cup \{\infty\}$ onto a closed polygon of $n+1$ sides, its sides are $\Delta_0, \Delta_1, \cdots, \Delta_n$ and its vertices are $w_0, w_1, \cdots, w_n, w_{n+1} = w_0$.

If $0 < \alpha_k < 2$, then we may prove that the inner angle at w_k is $\alpha_n\pi$.

We rewrite $\beta(t)$ as $\beta_k(t)(t - a_k)^{\alpha_k-1}$ where $\beta_k(t) = \prod_{j\neq k}(t - a_j)^{\alpha_j}$, $\beta_k(t)$ is holomorphic on a neighborhood V of a_k, we expand it as a Taylor series

$$\beta_k(t) = a_{0,k} + a_{1,k}(t - a_k) + \cdots, \qquad a_{0,k} \neq 0.$$

We have

$$f(z) = w_k + c\int_{a_k}^z (a_{0,k} + a_{1,k}(t - a_k) + \cdots)(t - a_k)^{\alpha_k-1}\, dt$$

$$= w_k + c\frac{a_{0,k}}{\alpha_k}(z - a_k)^{\alpha_k}\left(1 + \frac{\alpha_k}{\alpha_k + 1} \cdot \frac{a_{1,k}}{a_{0,k}}(z - a_k) + \cdots\right).$$

When z approaches to a_k along a straight line with inclination θ ($0 \leq \theta \leq \pi$), $f(z)$ approachs to w_k along a curve, and at the point w_k, the inclination of the tangent line is $\arg(a_{0,k}) + \alpha_k\theta$ since $c > 0$ and $\alpha_k > 0$. We construct a small semi-circle in $\overline{H}$ centered at a_k and contained in $V \cap \overline{H}$. When z varifies on the small semi-circle from $\theta = 0$ to $\theta = \pi$, $f(z)$ varifies on a Jordan curve from the point on Δ_{k-1} to the point on Δ_k, and the argument of $f(z)$ varifies from $\arg(a_{0,k})$ to $\arg(a_{0,k}) + \alpha_k\pi$. Hence the inner angle at w_k is $\alpha_k\pi$. The inner angle at the vertex $w_0 = w_{n+1}$ is $((n - 1) - (\alpha_1 + \cdots + \alpha_n))\pi > 0$, since the sum of the inner angles of a polygon of $n + 1$ sides equals to $(n - 1)\pi$. In particular, when $\alpha_1 + \cdots + \alpha_n = n - 2$, the inner angle at w_0 is π, that means, it is a polygon with n sides.

Formula (6.3) is Schwarz-Christoffel formula, where $\beta(t)$ is defined by (6.1), and $\alpha_1 + \cdots + \alpha_n < n - 1$. (6.3) maps the upper half plane to a polygon with $n+1$ sides, its vertices are $w_0, w_1, \cdots, w_n, w_{n+1} = w_0$, w_j $(j = 0, 1, \cdots, n+1)$ is defined by (6.2), its sides are $[w_{k-1}, w_k]$ $(k = 1, \cdots, n+1)$, the length of $[w_{k-1}, w_k]$ is l_k $(k = 1, \cdots, n+1)$. If $0 < \alpha_j < 2$ $(j = 1, \cdots, n)$, then the inner angle at w_k is $\alpha_k \pi$ $(k = 1, \cdots, n)$. The inner angle at $w_0 = w_{n+1}$ is $[(n-1) - (\alpha_1 + \cdots + \alpha_n)]\pi$.

We may write (6.3) in the more general form

$$f(z) = c \int_0^z \beta(t)\, \mathrm{d}t + c' \tag{6.3'}$$

if we write (6.2) in more general form

$$w_k = c \int_0^{a_k} (t - a_1)^{\alpha_1 - 1} \cdots (t - a_n)^{\alpha_n - 1}\, \mathrm{d}t + c', \qquad 0 \le k \le n+1,$$

where c, c' are two complex constants.

Formula (6.3') is Schwarz-Christoffel formula also.

EXERCISES IV

1. Verify each of the examples in §4.1.

2. Show that: Cross ratio is invariant under the linear fractional transformations.

3. Show that: If the entire function $f(z) = \sum\limits_{n=0}^{\infty} c_n z^n$ takes real value on the real axis, then all the coefficients c_n $(n = 0, 1, \cdots)$ are real numbers.

4. To construct the Riemann surface of $w = z + \sqrt{z^2 - 1}$.

5. Show that: any circle can be expressed as $\left|\frac{z - z_1}{z_1 - z_2}\right| = k$ $(k > 0)$ where z_1, z_2 are two symmetry points with respect to the circle, the center and the radius of the circle are

$$a = \frac{z_1 - k^2 z_2}{1 - k^2}, \qquad R = \frac{k|z_1 - z_2|}{|1 - k^2|}$$

respectively, when $k \ne 1$.

6. Show that: the cross ratio (z_1, z_2, z_3, z_4) is real if and only if z_1, z_2, z_3, z_4 are located at one circle.

7. (Carathéodory inequality) Using Schwarz lemma and the linear transformation to show that: If $f(z)$ is holomorphic on $|z| < R$, and continuous on

$|z| \leq R$, $M(r)$ and $A(r)$ denote the maximum values of $|f(z)|$ and $\mathrm{Re}\, f(z)$ on $|z| \leq R$ respectively, then

$$M(r) \leq \frac{2r}{R-r} A(R) + \frac{R+r}{R-r}|f(0)|$$

when $0 < r < R$.

8. If $f(z) = \sum\limits_{n=0}^{\infty} a_n z^n$ is holomorphic on the unit disk $|z| < 1$, and $|f(z)| \leq M$. Show that: $M|a_1| \leq M^2 - |a_0|^2$.

9. Construct each of the following conformal mappings.

(i) It maps the strip $0 < \mathrm{Im}\, z < \pi$ in the z-plane to $|w| < 1$;

(ii) It maps the semi-disk $|z| < 1$, $\mathrm{Im}\, z > 0$ in the z-plane to the upper half plane;

(iii) It maps the common part of $|z| < 1$ and $|z - 1| < 1$ in the z-plane to $|w| < 1$;

(iv) It maps the fan $\{z \mid 0 < \arg z < \alpha \ (< 2\pi), \ |z| < 1\}$ in the z-plane to $|w| < 1$;

(v) It maps the unit disk $|z| < 1$ in the z-plane to a strip $0 < v < 1$ ($w = u + iv$) in the w-plane, and maps $-1, 1, i$ to ∞, ∞, i respectively.

(vi) It maps the z-plane with a slit from $-\frac{1}{4}$ to $-\infty$ on the negative real axis to $|w| < 1$.

10. Using the Schwarz-Christoffel formula to give each of the conformal mappings which maps the upper half plane to the polygon P.

(i) P is a triangle, its vertices are w_1, w_2, w_3, the corresponding angles are $\alpha\pi, \beta\pi, \gamma\pi$ and $\alpha + \beta + \gamma = \pi$;

(ii) P is an equilateral triangle, its vertices are $w_1 = 0$, $w_2 = a$, $w_3 = \frac{1+\sqrt{3}i}{2} a$, where $a > 0$;

(iii) P is an isosceles triangle, its vertices are $w_1 = 0$, $w_2 = a$, $w_3 = a(1 + i)$, where $a > 0$;

(iv) P is a rectangular, its vertices are $w_1 = -k_1$, $w_2 = k_1$, $w_3 = k_1 + i k_2$, $w_4 = -k_1 + i k_2$, where $k_1 > 0$, $k_2 > 0$.

(v) P is a rhombus, its vertices are $0, a, a(1 + e^{i\alpha\pi}), ae^{i\alpha\pi}$, where $0 < \alpha < \frac{1}{2}$, $a > 0$;

(vi) P is a 5 sides regular polygon, its center is the origin, $w = 1$ is one vertex.

11. In the Riemann mapping theorem, if z_0 is real, U is a domain symmetric with respect to the real axis. Show that f satisfies the symmetric relation

$f(\bar{z}) = \overline{f(z)}$ by the uniqueness of the mapping.

12. Suppose $\mathcal{F}$ denotes the family of holomorphic function on a domain Ω and each function of the family takes the value on the right half plane. If there exists a point $z_0 \in \Omega$ such that $f(z_0) = g(z_0)$ for all $f \in \mathcal{F}$ and $g \in \mathcal{F}$. Show that $\mathcal{F}$ is a normal family.

13. Suppose $\mathcal{F}$ denotes the family of holomorphic functions on a domain Ω and each function of the family takes the value on $U_0 = \mathbb{C}\backslash\{x+i0, \ 0 \le x \le 1\}$. If there exists a point $z_0 \in \Omega$, such that $f(z_0) = g(z_0)$ for all $f \in \mathcal{F}$ and $g \in \mathcal{F}$. Show that $\mathcal{F}$ is a normal family.

APPENDIX   Riemann Surface

In §4.5 of this Chapter, we described the Riemann surface by examples. In this appendix, we will give the definition of Riemann surface.

Let X be a set, and let $\mathcal{F}$ be a family of the subsets of X, if (1) X and empty set belong to $\mathcal{F}$; (2) the sum of any elements in $\mathcal{F}$ belongs to $\mathcal{F}$ again; (3) the intersection of any two elements in $\mathcal{F}$ belongs to $\mathcal{F}$ again; then $\mathcal{F}$ is a topology of X, the elements of $\mathcal{F}$ are open sets. A set with topology is a **Topological space**, and is denoted by $(X, \mathcal{F})$. Roughly speaking, a topological space is a set when we define the open subset.

Let a set $U \subset X$, a point $x \in X$. If there exists an open set G in X so that $x \in G \subset U$, then U is a **neighborhood** of x.

A topological space is a Hausdoff space if for any two points in the space, we may find a neighborhood of each point, such that these two neighborhoods are non-interseted.

If X, Y are two topological spaces, and f is a one to one onto mapping from X to Y, and f and f^{-1} are continuous, then f is a **homeomorphism** of X to Y.

A Hausdorff space is a **surface** if it is locally homeomorphic to the Euclidean space. The meaning of a space is locally homeomorphic to Euclidean space is that for any point in the space there exists a neighborhood of this point which is homeomorphic to an open set in Euclidean space.

A **Riemann surface** is a connected Hausdorff topological space C, with a family of open covering $\{U_\alpha\}$ of C and a family of mappings $f_\alpha : U_\alpha \to C$ which satisfies the following conditions:

(a) $f : U_\alpha \to C$ is a homeomorphic mapping from U_α to an open set in C;

(b) If $U_\alpha \cap U_\beta \neq \emptyset$, the function

$$f_\beta \circ f_\alpha^{-1} : \ f_\alpha(U_\alpha \cap U_\beta) \longrightarrow f_\beta(U_\alpha \cap U_\beta)$$

is biholomorphic (i.e. the function and its inverse function are holomorphic). (U_α, f_α) is the locally holomorphic coordinates, $\{(U_\alpha, f_\alpha)\}$ is the covering of holomorphic coordinates.

Roughly speaking, a Riemann surface is a locally Euclidean Hausdorff space with complex structure.

The previous definition of Riemann surface is exactly the definition of complex manifold of dimension one. Similarly, we may define the complex manifold of high dimensions.

Now we give a simple example of Riemann surface.

Example   The extended complex plane $\mathbb{C}^* = \mathbb{C} \cup \{\infty\}$ is a Riemann surface.

It is easy to verify that $\mathbb{C}^*$ is a connected Hausdorff topological space.

Take the open covering $\{U_0, U_1\}$ of $\mathbb{C}^*$, where

$$U_0 = \mathbb{C}, \qquad U_1 = \mathbb{C}^* \backslash \{0\},$$

and let $f_0(z) = z$, $f_1(z) = 0$ if $z = \infty$; $f_1(z) = \frac{1}{z}$ if $z \neq \infty$, then $f_1 \circ f_0^{-1}$ and $f_0 \circ f_1^{-1}$ are biholomorphic on $\mathbb{C} \backslash \{0\}$.

Thus, $\mathbb{C}^*$ is a Riemann surface. It is easy to verify the Riemann sphere is a Riemann surface.

The further discussion about the Riemann surface, for example, how to define differential, integral on Riemann surface, how to classify the Riemann surfaces, how to define the meromorphic functions, zero point and pole on Riemann surface, all these topics are out of this textbook, the readers may read the references if they are interested.

CHAPTER V
DIFFERENTIAL GEOMETRY
AND PICARD THEOREM

§ 5.1   Metric and Curvature

In this chapter, we will introduce the elementary facts about complex differential geometry, and use it to treat some theorems on the complex analysis, for example, Picard theorem. Picard theorem is one of the most important classical theorems in complex analysis, especially in the theory of value distribution. The original proof of Picard theorem is complicated. Now we use differential geometry to prove it, the proof is clear and simple.

Let Ω be a domain in $\mathbb{C}$. A metric on Ω is a non-negative C^2 function ρ on Ω, $\mathrm{d}s_\rho^2 = \rho^2|\mathrm{d}z|^2$. This metric give the distance function d. The distance between two points $z_1 \in \Omega$ and $z_2 \in \Omega$ is defined by

$$d(z_1, z_2) = \inf \int_\gamma \rho(z)|\mathrm{d}z|, \tag{1.1}$$

where infimum are taking from all curves γ which connect z_1 and z_2, and inside Ω.

For metric ρ, we define the curvature

$$K(z, \rho) = -\frac{\Delta \log \rho(z)}{\rho^2(z)} \tag{1.2}$$

when $\rho(z) \neq 0$, where Δ is the Laplace operator,

$$\Delta = \frac{\partial^2}{\partial x^2} + \frac{\partial^2}{\partial y^2} = 4\frac{\partial}{\partial z}\frac{\partial}{\partial \overline{z}} = 4\frac{\partial}{\partial \overline{z}}\frac{\partial}{\partial z} = \frac{\partial^2}{\partial r^2} + \frac{1}{r}\frac{\partial}{\partial r} + \frac{1}{r^2}\frac{\partial^2}{\partial \theta^2},$$

here $z = x + \mathrm{i}\,y = r e^{\mathrm{i}\,\theta}$.

We may prove that the curvature which is defined by (1.2) is the Gauss curvature in differential geometry (cf. Appendix of this Chapter).

In complex differential geometry, usually we use the following three important metrics.

1. Euclidean metric.

If $\Omega = \mathbb{C}$, Ω equips with a metric $\rho(z) = 1$. That means, for any point $z \in \mathbb{C}$, $ds^2 = |dz|^2$. This metric is **Euclidean metric** or **parabolic metric**. The distance between two points z_1 and z_2 is the Euclidean distance, $d(z_1, z_2) = \inf \int_\gamma |dz| = |z_1 - z_2| = $ the length of the line segment which connect z_1 and z_2.

The transformation $w = e^{i\theta} z + a$ is composed by a rotation $w = e^{i\theta} z$ and a translation $w = z + a$. All these transformations $\{w = e^{i\theta} z + a \mid \theta$ is any real number, a is any complex number$\}$ form a group. It is the group of Euclidean motions, or the group of rigid motions. It is a subgroup of Aut $(\mathbb{C})$. Obviously, Euclidean metric is invariant under the group of Euclidean motions. For this metric, we have $K(z, \rho) = 0$ for any $z \in \mathbb{C}$ by (1.2). This is why we call it parabolic metric.

2. Poincaré metric.

If Ω is the unit disk $D(0,1) = \{z \mid |z| < 1\}$, Ω equips with a metric $\rho = \lambda(z) = \frac{2}{1-|z|^2}$, $ds_\lambda^2 = \frac{4|dz|^2}{(1-|z|^2)^2}$. This metric is **Poincaré metric** or **hyperbolic metric**. The transformation $w = e^{i\theta} \frac{z-a}{1-\bar{a}z}$, where θ is any real number and $a \in D(0,1)$, is composed by a rotation and a Möbius transformation. In Chapter II §2.5, we already proved that all the transformations $\{w = e^{i\theta} \frac{z-a}{1-\bar{a}z} \mid \theta$ is any real number, $a \in D(0,1)\}$ form the group of holomorphic automorphisms Aut $(D(0,1))$ of $D(0,1)$. In Chapter II §2.5, we already proved that Poincaré metric is invariant under Aut $(D(0,1))$.

Now we try to evaluate the Poincaré distance between two points z_1 and z_2 in $D(0,1)$.

We consider the Poincaré distance between points $z_1 = 0$ and $z_2 = R + i0$ $(R < 1)$ at first. The curve γ connecting these two points can be written as
$$z(t) = u(t) + i\,w(t), \qquad 0 \le t \le 1,$$
$$w(0) = u(0) = w(1) = 0, \qquad u(1) = R,$$
and $u(t)^2 + w(t)^2 < 1$, u, w are real valued C^2 functions of t. We have
$$\int_\gamma ds = \int_\gamma \frac{2|dz|}{1-|z|^2} = 2 \int_0^1 \frac{((u'(t))^2 + (w'(t))^2)^{\frac{1}{2}} \, dt}{1 - (u(t))^2 - (w(t))^2}$$
$$\ge \int_0^1 \frac{2|u'(t)|dt}{1 - (u(t))^2} \ge \left| \int_0^R \frac{2du}{1-u^2} \right| = \log \frac{1+R}{1-R},$$

equality holds if and only if $w(t) = 0$, $0 \le t \le 1$. Hence

$$d(0, R + i0) = \inf_\gamma \int_\gamma \frac{|dz|}{1 - |z|^2} = \log \frac{1 + R}{1 - R}$$

and the γ which makes the infimum of the integral is the straight line segment connecting 0 and $R + i0$.

The transformation $w = e^{i\theta} z$ is an element of Aut $(D(0, 1))$. The Poincaré distance between any two points in $D(0, 1)$ is invariant under the transformation $w = e^{i\theta} z$. Thus

$$d(0, R e^{i\theta}) = \log \frac{1 + R}{1 - R}$$

holds for any real number θ.

If z_1, z_2 are any two points in $D(0, 1)$, then

$$\varphi(z) = \frac{z - z_1}{1 - \overline{z}_1 z}$$

is an element of Aut $(D(0, 1))$, which maps z_1 to 0, z_2 to $\frac{z_2 - z_1}{1 - \overline{z}_1 z_2}$. Thus

$$d(z_1, z_2) = d\left(0, \frac{z_2 - z_1}{1 - \overline{z}_1 z_2}\right) = \log \frac{1 + \left|\dfrac{z_2 - z_1}{1 - \overline{z}_1 z_2}\right|}{1 - \left|\dfrac{z_2 - z_1}{1 - \overline{z}_1 z_2}\right|}. \qquad (1.3)$$

This is the Poincaré distance or hyperbolic distance between two points z_1, z_2 in $D(0, 1)$. We know that

$$d(z_1, z_2) = \inf_\gamma \int_\gamma \frac{|dz|}{1 - |z|^2}.$$

The curve γ which makes the infimum of the integral is the curve

$$z = \frac{z_1 + \dfrac{z_2 - z_1}{1 - \overline{z}_1 z_2} t}{1 + \overline{z}_1 \dfrac{z_2 - z_1}{1 - \overline{z}_1 z_2} t}, \qquad 0 \le t \le 1,$$

i.e.,

$$z = \frac{(1 - t)z_1 + (t - z_1 \overline{z}_1) z_2}{1 - t z_1 \overline{z}_1 - (1 - t) \overline{z}_1 z_2}, \qquad 0 \le t \le 1.$$

From (1.3), we find that $d(z_1, z_2) = 0$ when $z_2 \to z_1$; and $d(z_1, z_2) \to \infty$ when z_1 or z_2 approaches to some boundary point of $D(0, 1)$.

Theorem 19 of Chapter II §2.5 is Schwarz-Pick lemma: If $w = f(z)$ is a holomorphic function on $D(0, 1)$, f maps $D(0, 1)$ into $D(0, 1)$, and $w_1 = f(z_1)$, $w_2 = f(z_2)$, then

$$\left| \frac{w_1 - w_2}{1 - \overline{w}_1 w_2} \right| \le \left| \frac{z_1 - z_2}{1 - \overline{z}_1 z_2} \right| \tag{1.4}$$

equality holds if and only if $f(z) \in \mathrm{Aut}\,(D(0, 1))$.

By (1.3), we may rewrite (1.4) as

$$d(w_1, w_2) \le d(z_1, z_2).$$

Thus the Schwarz-Pick lemma has a clear geometry meaning: If $w = f(z)$ is a holomorphic function on $D(0, 1)$, f maps $D(0, 1)$ into $D(0, 1)$, then the Poincaré distance between any two points in $D(0, 1)$ is non-increasing after this mapping. The distance is invariant if and only if $f(z) \in \mathrm{Aut}\,(D(0, 1))$.

Using $\Delta = 4 \frac{\partial}{\partial z} \frac{\partial}{\partial \overline{z}}$, we have

$$-\Delta \log \lambda(z) = \Delta \log(1 - |z|^2) = \frac{-4}{(1 - |z|^2)^2},$$

and hence the curvature of hyperbolic metric $K(z, \lambda) = -1$ for all $z \in D(0, 1)$. This is why we call this metric hyperbolic metric.

3. Spherical metric.

If Ω is $\mathbb{C}^*$, C^* equips with a metric $\rho = \sigma(z) = \frac{2}{1 + |z|^2}$,

$$ds_\sigma^2 = \frac{4|dz|^2}{(1 + |z|^2)^2}.$$

This metric is **spherical metric** or **elliptic metric**. In Chapter I §1.2, we discussed the stereographic projection which established the one to one correspondence between the Riemann sphere S^2 and $\mathbb{C}^*$. If $z \in \mathbb{C}^*$, the coordinates of the corresponding point of z in S^2 are

$$(x_1, x_2, x_3) = \left\{ \frac{z + \overline{z}}{|z|^2 + 1}, \frac{z - \overline{z}}{i\,(|z|^2 + 1)}, \frac{|z|^2 - 1}{|z|^2 + 1} \right\}. \tag{1.5}$$

If $P = (x_1, x_2, x_3)$ and $P' = (x_1', x_2', x_3')$ are two points on S^2, the shortest distance between P and P' is attained by some great circular arc $\overset{\frown}{PP'}$. It is easy to verify that the arc length equals to

$$2\arctan\sqrt{\frac{1 - x_1 x_1' - x_2 x_2' - x_3 x_3'}{1 + x_1 x_1' + x_2 x_2' + x_3 x_3'}},$$

which is 2 arc $\tan\left|\frac{z-z'}{1+z\bar{z}'}\right|$ by (1.5). Using this distance as the distance between two corresponding points z, z' in $\mathbb{C}^*$, we have

$$d(z, z') = 2\mathrm{arc\ tan}\left|\frac{z - z'}{1 + z\bar{z}'}\right|.$$

Differentiating $d(z, z')$, we get the corresponding metric

$$\mathrm{d}s^2 = \sigma(z)^2|\mathrm{d}z|^2, \qquad \sigma(z) = \frac{2}{1 + |z|^2}.$$

Thus the metric $\sigma(z)$ has a concrete geometric meaning. If we use the metric $\sigma(z)$ to evaluate the distance between two points in $\mathbb{C}^*$, it equals to the shortest distance between two corresponding points in S^2, that is the spherical distance. In other words, we use the spherical distance between two corresponding points on S^2 as the distance between two points in $\mathbb{C}^*$. We have

$$d(z_1, z_2) = \inf_\gamma \int_\gamma \sigma(z)|\mathrm{d}z|,$$

where γ is any curve connecting z_1 and z_2.

If the corresponding points on S^2 of z_1, z_2 are P_1 and P_2, the stereographic projection projects the great circular arc $\overset{\frown}{P_1 P_2}$ to a curve γ_0 in $\mathbb{C}^*$ connecting z_1 and z_2, then γ_0 is the curve which makes infimum of the above integral. This is the reason why we call the metric $\sigma(z)$ spherical metric.

It is easy to verify that the curvature of spherical metric $\sigma(z)$ is $+1$ at every point $z \in \mathbb{C}^*$. This is why we call this metric elliptic metric.

In Chapter III §3.3, we stated the important uniformization theorem: Any simply connected Riemann surface is one to one holomorphic equivalent to one and only one of the following three domains: $\mathbb{C}$; $D(0, 1)$ and $\mathbb{C}^*$. This is the reason why we study the geometry of these three domains.

The curvature was defined by (1.2). It has the following important property: curvature is invariant under holomorphic mappings.

If Ω_1, Ω_2 are two domains, and a holomorphic function f maps Ω_1 onto Ω_2. If ρ is a metric on Ω_2, and $f' \not\equiv 0$, then

$$f^*\rho = (\rho \circ f)|f'| \tag{1.6}$$

defines a metric on Ω_1. We call this metric on Ω_1 as the pullback of the metric ρ on Ω_2 under the map f. We need to prove

$$K(z, f^*\rho) = K(f(z), \rho). \tag{1.7}$$

It is easy to verify

$$\Delta \log |f'(z)| = 0$$

and

$$\begin{aligned}
\Delta \log (\rho \circ f(z)) &= 4\frac{\partial}{\partial \bar{z}}\frac{\partial}{\partial z} \log (\rho \circ f(z)) \\
&= 4\overline{\left(\frac{\partial f}{\partial z}\right)}\left(\frac{\partial f}{\partial z}\right)\left(\frac{\partial}{\partial f}\frac{\partial}{\partial \bar{f}} \log (\rho \circ f)\right) \\
&= |f'(z)|^2(\Delta_f \log \rho) \circ f(z).
\end{aligned}$$

Thus

$$\begin{aligned}
K(z, f^*\rho) &= -\frac{|f'(z)|^2(\Delta_f \log \rho) \circ f(z)}{(\rho \circ f(z))^2|f'(z)|^2} \\
&= -\frac{(\Delta_f \log \rho) \circ f(z)}{(\rho \circ f(z))^2} \\
&= K(f(z), \rho).
\end{aligned}$$

§ 5.2   Ahlfors-Schwarz Lemma

The Theorem 17 of Chapter II §2.5 is the analytic form of the classical Schwarz lemma. The Theorem 19 of Chapter II §2.5 is the Schwarz-Pick lemma which is an extension of the classical Schwarz lemma. In previous section, we gave the differential geometry interpretation of Schwarz-Pick lemma which was described by Poincaré metric. Here we will give another extension of the classical Schwarz lemma, namely, the Ahlfors-Schwarz lemma, which is described by curvature and is an extension of Schwarz-Pick lemma. The Ahlfors-Schwarz lemma was established by Ahlfors at 1938 (L.V. Ahlfors[5]). This lemma marks the beginning of the differential geometry entering complex analysis, and is the start point to treat complex analysis by differential geometry.

Theorem 1 (Ahlfors-Schwarz lemma)   Suppose $f(z)$ is holomorphic on $D(0,1)$, and f maps $D(0,1)$ onto U. If U equips with a metric $\rho(z)$, $ds_\rho^2 = \rho(z)^2 |dz|^2$, whose curvature is ≤ -1 on every point in U, then

$$f^* \rho(z) \leq \lambda(z), \tag{2.1}$$

that is,

$$ds_\rho^2 \leq ds_\lambda^2, \tag{2.2}$$

where $\lambda(z) = \frac{2}{1-|z|^2}$. That means: the metric is non-increasing after mapping.

Proof   Fixing $r \in (0,1)$, we define a metric on the disk $D(0,r)$ (a disk centered at 0, with radius r) as

$$\lambda_r(z) = \frac{2r}{r^2 - z^2}.$$

Obviously, the curvature of this metric on any $z \in D(0,r)$ equals -1. Define the function

$$v(z) = \frac{f^* \rho(z)}{\lambda_r}$$

on $D(0,r)$, then $v(z)$ is a non-negative continuous function. By (1.6), $f^*\rho(z) = \rho(f(z))|f'(z)|$ is bounded on $\overline{D(0,r)}$. Observing $\frac{1}{\lambda_r} \to 0$ as $|z| \to r$, we have $v \to 0$ as $|z| \to r$. Thus v attains its maximum value M at some inner point τ in $D(0,r)$. If we can prove that $M \leq 1$, then $v \leq 1$ on $D(0,r)$. Let $r \to 1^-$, we have (2.1).

If $f^*\rho(\tau) = 0$, then $v \equiv 0$. We need not to prove anymore. Suppose $f^*\rho(\tau) > 0$, then $K(\tau, f^*\rho)$ makes sense. By the assumption $K(\tau, f^*\rho) \leq -1$, and the fact that $\log v$ takes its maximum value at τ, we have

$$\begin{aligned} 0 \geq & \Delta \log v(z) = \Delta \log f^* \rho(\tau) - \Delta \log \lambda_r(\tau) \\ = & -K(\tau, f^*\rho) \cdot (f^*\rho(\tau))^2 + K(\tau, \lambda_r)(\lambda_r(\tau))^2 \\ \geq & (f^*\rho(\tau))^2 - (\lambda_r(\tau))^2, \end{aligned}$$

that is,

$$\frac{f^*\rho(\tau)}{\lambda_r(\tau)} \leq 1.$$

Thus $M \leq 1$. The lemma have proved.

If $U \subseteq D(0,1)$ in Ahlfors-Schwarz lemma, we may take $\rho = \lambda$, then we get Schwarz-Pick lemma. Hence, Ahlfors-Schwarz lemma is an extension of Schwarz-Pick lemma.

We may extend the Ahlfor-Schwarz lemma to a more general form.

We define a metric

$$\lambda_\alpha^A(z) = \frac{2\alpha}{\sqrt{A}\,(\alpha^2 - |z|^2)} \tag{2.3}$$

on $D(0,\alpha)$, where $\alpha > 0$, $A > 0$. The curvature of this metric equals $-A$ at any point in $D(0,\alpha)$.

Theorem 2 (General form of Ahlfors-Schwarz lemma) Suppose $f(z)$ is holomorphic on $D(0,\alpha)$, and f maps $D(0,\alpha)$ to U. If U equips with a metric ρ, $\mathrm{d}s_\rho^2 = \rho(z)^2|\mathrm{d}z|^2$, whose curvature is not greater than $-B$ on any point in U, then

$$f^*\rho(z) \le \frac{\sqrt{A}}{\sqrt{B}}\,\lambda_\alpha^A(z) \tag{2.4}$$

holds for every $z \in D(0,\alpha)$, where B is a positive constant.

The proof of Theorem 2 is similar with Theorem 1, we omit the detail of the proof.

Ahlfors-Schwarz lemma is one start point of the comparison theorem in differential geometry.

Using Ahlfors-Schwarz lemma, we may get many important results, for example, the extension of Liouville theorem, etc.

§ 5.3   Extension of Liouville Theorem and Value Distribution

Theorem 8 of Chapter II §2.3 is the important Liouville theorem: Any bounded entire function is a constant. Now we use Ahlfors-Schwarz lemma to extend the Liouville theorem and to describe it by curvature.

Theorem 3 (Extension of Liouville theorem)   If the entire function $f(z)$ maps $\mathbb{C}$ onto U, and U admits a metric $\rho(z)$ whose curvature $K(z,\rho)$ satisfies $K(z,\rho) \le -B < 0$ for every $z \in U$, where B is a positive constant, then $f(z)$ is a constant.

Proof   For any $\alpha > 0$, $f(z)$ maps $D(0,\alpha)$ into U. By assumption, U admits a metric $\rho(z)$ whose curvature $K(z,\rho)$ satisfies $K(z,\rho) \le -B < 0$. We have

$$f^*\rho(z) \leq \frac{\sqrt{A}}{\sqrt{B}}\, \lambda_\alpha^A(z)$$

by Theorem 2. From (2.3), we know that $\lambda_\alpha^A(z) \to 0$ when $\alpha \to \infty$. Thus $f^*\rho(z) \leq 0$. Hence $f^*\rho(z) = 0$. It implies that f is a constant since $f(z)$ is holomorphic. We have proved the lemma.

Theorem 3 implies Liouville theorem.

If $f(z)$ is a bounded entire function, then there exist a positive constant M such that $|f(z)| \leq M$ holds for all $z \in \mathbb{C}$. The holomorphic function $\frac{1}{M}f(z)$ maps $\mathbb{C}$ into $D(0,1)$. Since $D(0,1)$ admits a metric λ whose curvature is -1, let $B = 1$ in Theorem 3, then $\frac{1}{M}f(z)$ is a constant. Hence $f(z)$ is a constant. It proved the Liouville theorem.

Thus Theorem 3 is the extension of Liouville theorem in the form of differential geometry.

From Liouville theorem, we know that: If the entire function $w = f(z)$ maps $\mathbb{C}$ onto a bounded domain, then $f(z)$ is a constant. If the entire function $w = f(z)$ maps $\mathbb{C}$ onto an unbounded domain U, and the area of $\mathbb{C}\backslash U$ is > 0, then we may still prove that $f(z)$ is a constant. The proof is as follows. If $c_1 \in \mathbb{C}\backslash U$, and c_1 is an inner point of $\mathbb{C}\backslash U$, then $w_1 = 0$ is located in the complement of $f(\mathbb{C}) - c_1$, where $w_1 = w - c_1$. The transformation $w_2 = \frac{1}{w-c_1}$ maps $\mathbb{C}$ onto a bounded domain. Hence w_2 is a constant, so is w.

Moreover, if the entire function $w = f(z)$ maps $\mathbb{C}$ onto an unbounded domain U, and the area of $\mathbb{C}\backslash U$ is zero, it means that $\mathbb{C}\backslash U$ is composed by some curves. We may ask that does $f(z)$ a constant?

We observe the following example.

If the entire function $w = u + iv = f(z)$ maps $\mathbb{C}$ onto $\mathbb{C}\backslash\{u + i0 \mid 0 \leq u \leq 1\}$. The transformation

$$w_1 = u_1 + i\,v_1 = \varphi(w) = \frac{w}{w-1}$$

maps $\mathbb{C}$ onto $\mathbb{C}\backslash\{u_1 + i0 \mid u_1 \leq 0\}$. The transformation $w_2 = r(w_1) = \sqrt{w_1}$, where we take the principal branch of square root, maps $\mathbb{C}$ onto the right upper half plane. The Cayley transformation $w_3 = \frac{w_2-1}{w_2+1} = s(w_2)$ maps the right upper half plane to the unit circle. Then w_3 is a constant by Liouville theorem, and hence w_2, w_1 and w are constants.

From this example, we find that even the entire function $w = f(z)$ maps $\mathbb{C}$ onto a unbounded domain U, and $\mathbb{C}\backslash U$ is a line segment only, the entire function is a constant again. Of course, even the length of the line segment is very small, the entire function is a constant again.

How small size of $\mathbb{C}\backslash U$ can imply that $f(z)$ is not a constant function ?

An extreme example is the entire function $f(z) = e^z$ which maps $\mathbb{C}$ onto $U = \mathbb{C}\backslash\{0\}$. It means that if $\mathbb{C}\backslash U$ is one point, then we have an example to show that $f(z)$ is not a constant. If $\mathbb{C}\backslash U$ has two points, does $f(z)$ a constant ? The answer is the Picard little theorem.

§ 5.4   Picard Little Theorem

Theorem 4 (Picard little theorem)   If the entire function $w = f(z)$ maps $\mathbb{C}$ onto U, and $\mathbb{C}\backslash U$ contains two points at least, then $f(z)$ is a constant.

That is: The image of a non-constant entire function contains all complex numbers with the possible exception of one value.

In purpose to prove Picard little theorem, we need to prove the following theorem.

Theorem 5   If U is an open set in $\mathbb{C}$, and $\mathbb{C}\backslash U$ contains two points at least, then U admits a metric μ whose curvature $K(z,\mu)$ satisfies

$$K(z,\mu) \leq -B < 0$$

on any point in U, where B is a positive constant.

Theorem 4 is a consequence of Theorem 5.

If the entire function $w = f(z)$ maps $\mathbb{C}$ onto U, and $\mathbb{C}\backslash U$ contains two points at least, then Theorem 5 implies that U admits a metric μ, whose curvature satisfies $K(z,\mu) \in -B < 0$ on U, where B is a positive constant. By the extension of Liouville theorem (Theorem 3), $f(z)$ is a constant.

Proof of Theorem 5   We may transform two points in $\mathbb{C}\backslash U$ onto 0 and 1 by a linear transformation. Let $\mathbb{C}_{0,1} = \mathbb{C}\backslash\{0,1\}$. Equip with a metric

$$\mu(z) = \frac{(1+|z|^{\frac{1}{3}})^{\frac{1}{2}}}{|z|^{\frac{5}{6}}} \cdot \frac{(1+|z-1|^{\frac{1}{3}})^{\frac{1}{2}}}{|z-1|^{\frac{5}{6}}} \tag{4.1}$$

on $\mathbb{C}_{0,1}$, then $\mu(z)$ is a positive smooth function on $\mathbb{C}_{0,1}$. Now we evaluate the curvature of μ and to prove it is negative.

Obviously

$$\Delta(\log|z|^{\frac{5}{6}}) = \frac{5}{12}\Delta(\log|z|^2) = 0,$$

we have

$$\Delta \log \frac{(1+|z|^{\frac{1}{3}})^{\frac{1}{2}}}{|z|^{\frac{5}{6}}} = \frac{1}{2}\Delta \log (1+|z|^{\frac{1}{3}})$$

$$=2\frac{\partial}{\partial z}\frac{\partial}{\partial \overline{z}} \log(1+(z\cdot\overline{z})^{\frac{1}{6}}) = \frac{1}{18|z|^{\frac{5}{3}}(1+|z|^{\frac{1}{3}})^2}.$$

Similarly, we have

$$\Delta \log \left[\frac{(1+|z-1|^{\frac{1}{3}})^{\frac{1}{2}}}{|z-1|^{\frac{5}{6}}}\right] = \frac{1}{18|z-1|^{\frac{5}{3}}(1+|z-1|^{\frac{1}{3}})^2}.$$

Then the curvature

$$K(z,\mu) = -\frac{1}{18}\left[\frac{|z-1|^{\frac{5}{3}}}{(1+|z|^{\frac{1}{3}})^3(1+|z-1|^{\frac{1}{3}})} + \frac{|z|^{\frac{5}{3}}}{(1+|z|^{\frac{1}{3}})(1+|z-1|^{\frac{1}{3}})^3}\right].$$

It is easy to verify that

(1) $K(z,\mu) < 0$ for all $z \in \mathbb{C}_{0,1}$;

(2) $\lim_{z\to 0} K(z,\mu) = -\frac{1}{36}$;

(3) $\lim_{z\to 1} K(z,\mu) = -\frac{1}{36}$;

(4) $\lim_{z\to\infty} K(z,\mu) = -\infty$.

Thus there exists a negative constant $-B$ as the upper bound of $K(z,\mu)$ on $\mathbb{C}_{0,1}$.

We have proved Theorem 5.

We will prove a more deep theorem, the Picard great theorem. It is the deepening of the Picard little theorem. In purpose to prove the Picard great theorem, we need to extend the idea of normal family.

§ 5.5   Extension of Normal Family

We established the idea of normal family at Chapter IV §4.2, and used it to prove the Riemann mapping theorem. Now we extend this idea.

Definition 1   If $\{g_j\}$ is a sequence of complex-valued functions (may not holomorphic) on domain Ω, for any given $\varepsilon > 0$ and any compact subset K in Ω, there exists a positive integer J depending on ε and K only, and a function $g(z)$ on Ω, such that

$$|g_j(z) - g(z)| < \varepsilon$$

holds for any $z \in K$ when $j > J$, then $\{g_j\}$ is **normally convergent** to $g(z)$ on Ω.

In other words, $\{g_j\}$ is normally convergent on Ω if $\{g_j\}$ converges uniformly on any compact subset of Ω.

If for any compact subset K in Ω, and any compact set L in $\mathbb{C}$, there exists a positive integer J depending on K and L only, such that $g_j(z) \notin L$ for any $z \in K$ when $j > J$, then $\{g_j\}$ is **compactly divergent** on Ω.

In other words, $\{g_j\}$ is compact divergent if $\{g_j\}$ diverges uniformly to ∞ at any compact subset in Ω.

Definition 2 If $\mathcal{F}$ is a family of complex-valued functions on domain Ω, for any sequence, there exists a subsequence either normally convergent or compactly divergent on Ω, then $\mathcal{F}$ is a **normal family** on Ω.

This is the extension of the idea of normal family which was defined at Chapter IV §4.2.

Example $\mathcal{F} = \{f_j\}$, $f_j = z^j$, $j = 1, 2, \cdots$, $\mathcal{F}$ is a normal family on $D(0,1)$, because any subsequence is normally convergent to zero on $D(0,1)$. $\mathcal{F}$ is a normal family on $\{z \mid |z| > 1\}$, because any subsequence is compact divergent on $\{z \mid |z| > 1\}$. $\mathcal{F}$ is not a normal family on any domain containing the unit circle $|z| = 1$ as its inner points, because any subsequence is convergence on the point inside the circle, and divergence on the point outside the circle.

From this definition, we have the following theorem.

Theorem 6 (Montel theorem) If $\mathcal{F}$ is a family of holomorphic functions on Ω, for any compact subset K in Ω, there exists a constant M_K, such that

$$|f(z)| \leq M_K \tag{5.1}$$

holds for any $z \in K$, $f \in \mathcal{F}$, then $\mathcal{F}$ is a normal family.

If $|f(z)| \leq M$ holds for every $z \in \Omega$, $f \in \mathcal{F}$, where M is a constant, then the theorem holds true again.

Since $\mathcal{F}$ is a family of holomorphic functions and satisfies the condition (5.1), the subsequence is impossible compactly divergent. By Theorem 3 (Montel Theorem) of Chapter IV §4.2, we know Theorem 6 is true.

In purpose to extend the idea of normal family to the family of meromorphic functions, we use the spherical distance in S^2 to replace the Euclidean distance in $\mathbb{C}$. We may define the normal family of meromorphic functions on $\mathbb{C}^*$ as follows.

Definition 3   If $\mathcal{F}$ is a family of meromorphic functions on a domain $\Omega \subset \mathbb{C}^*$, for any sequence there exists a subsequence which is normally convergent on Ω in the sense of spherical distance, then $\mathcal{F}$ is a normal family.

The form of Definition 3 is the same as the definition of normal family (Definition 1) in Chapter IV §4.2, if we replace the Euclidean distance by the spherical distance.

The Definition 2 and Definition 3 are consistent if we replace Euclidean distance by spherical distance.

Similar with Montel theorem (Theorem 6), we have the following Marty theorem.

Theorem 7 (Marty theorem)   If $\mathcal{F}$ is a family of meromorphic functions on Ω, then $\mathcal{F}$ is a normal family if and only if

$$\{f^*\sigma \mid f \in \mathcal{F}\} \tag{5.2}$$

is equibounded on any compact subset in Ω, where σ is the spherical metric, that is, for any compact subset K in Ω, there exists a constant M_K, such that

$$\frac{2|f'(z)|}{1+|f(z)|^2} \leq M_K \tag{5.3}$$

holds for any $z \in K$, $f \in \mathcal{F}$.

Proof   (5.2) is equibounded on any compact subset in Ω is equivalent to (5.3).

If (5.3) holds, then

$$d(f(z_1), f(z_2)) = \inf \int_\gamma \mathrm{d}s = \inf \int_{\gamma'} \frac{2|f'(z)|}{1+|f(z)|^2} |\mathrm{d}z|$$
$$\leq \int_{\gamma_0} \frac{2|f'(z)|}{1+|f(z)|^2} |\mathrm{d}z| \leq M_K |z_1 - z_2|,$$

where γ is the curve inside K connecting $f(z_1)$ and $f(z_2)$, γ' is $f^{-1}(\gamma)$, and γ_0 is the line segment connecting z_1 and z_2. Hence, $\mathcal{F}$ is equicontinuous in the sense of spherical distance. And $\mathcal{F}$ is equibounded since the spherical distance is finite. Thus $\mathcal{F}$ is a normal family, by Ascoli-Arzela theorem (Chapter IV §4.2 Theorem 4).

Conversely, if $\mathcal{F}$ is a normal family, we need to prove (5.3) holds true. If (5.3) is not true, then there exists a compact subset E and a sequence $\{f_n\}$ in

$\mathcal{F}$, such that $\max\limits_{z\in E} f_n^*\sigma(z)$ is unbounded. Since $\mathcal{F}$ is a normal family, there exists a subsequence $\{f_{n_k}\}$ in $\{f_n\}$, so that $f_{n_k}(z) \to f(z)$ holds on E when $n_k \to \infty$. For any point in E, we may find a closed disk $\overline{D}$, $\overline{D} \subset \Omega$, and f is holomorphic or $\frac{1}{f}$ is holomorphic on $\overline{D}$. If f is holomorphic, then f is bounded on $\overline{D}$. Since $\{f_{n_k}\}$ is convergence in the sense of spherical distance, $\{f_{n_k}\}$ has no pole in $\overline{D}$ when n_k is sufficiently large. By Chapter III §3.1 Theorem 1 (Weierstrass theorem), $f_{n_k}^*\sigma$ converges uniformly to $f^*\sigma$ on a disc which is inside $\overline{D}$. Since $f^*\sigma$ is a continuous function, $f_{n_k}^*\sigma$ is bounded on this disc. Similarly, if $\frac{1}{f}$ is holomorphic, we may prove that $(\frac{1}{f_{n_k}})^*\sigma$ is bounded on a disc which is inside $\overline{D}$. Hence $f_{n_k}^*\sigma$ is bounded on this disc since $(\frac{1}{f_{n_k}})^*\sigma = f_{n_k}^*\sigma$. Since E is a compact subset, we may find a finite number of small discs to cover it. Thus $f_{n_k}^*\sigma$ is bounded on E. We obtain a contradiction.

From Marty theorem, we have the following Montel theorem.

Theorem 8 (Montel theorem) If $\mathcal{F}$ is a family of meromorphic functions on domain Ω, P, Q, R are three different points, and any function in $\mathcal{F}$ assumes the value $\mathbb{C}^*\backslash\{P, Q, R\}$, then $\mathcal{F}$ is a normal family.

Proof We may transform P, Q, R to 0, 1, ∞ respectively by a linear factional transformation. We need only to prove that any function in a family of holomorphic functions does not assume the value 0 and 1, then the family is a normal family. That means if the family of holomorphic functions which assume the values on $\mathbb{C}_{0,1} = \mathbb{C}\backslash\{0,1\}$, then the family is a normal family. We need to prove that $\mathcal{F}$ is a normal family on any disc $D(z_0, \alpha) = \{z \mid |z - z_0| < \alpha\} \subseteq \Omega$. Without loss of generality, we may assume that $z_0 = 0$. In §5.4, we already constructed a metric μ on $\mathbb{C}_{0,1}$, whose curvature ≤ -1 (we may multiply a constant a on μ if it is needed). By the general form of Ahlfors-Schwarz lemma (Theorem 2) in §5.2, we know that for any $f \in \mathcal{F}$, we have

$$f^*\mu(z) \leq \lambda_\alpha^A(z),$$

it is

$$\mu(f(z))\left|\frac{df}{dz}\right| \leq \frac{2\alpha}{\sqrt{A}\,(\alpha^2 - |z|^2)} \tag{5.4}$$

holds for every $z \in D(0, \alpha)$.

Comparing the spherical metric $\sigma(w)$ and $\mu(w)$ on $\mathbb{C}_{0,1}$, we find that

$$\frac{\sigma(w)}{\mu(w)} = \frac{2/(1+|w|^2)}{a(1+|w|^{1/3})^{1/2}(1+|w-1|^{1/3})^{1/2}/[|w|^{5/6}|w-1|^{5/6}]} \longrightarrow 0$$

when $w \to 0$ or $w \to 1$ or $w \to \infty$. Hence, there exists a positive constant M such that $\sigma(w) \le M\mu(w)$. By (5.4)

$$f^*\sigma(z) = \sigma(f(z))\left|\frac{\mathrm{d}f}{\mathrm{d}z}\right| \le M\mu(f(z))\left|\frac{\mathrm{d}f}{\mathrm{d}z}\right|$$

$$= Mf^*\mu(z) \le M\lambda_\alpha^A = \frac{2\alpha M}{\sqrt{A}\,(\alpha^2 - |z|^2)}$$

when $z \in D(0, \alpha)$. Hence $f^*\sigma$ is bounded on compact subsets in $D(0, \alpha)$, and the upper bound is independent of $f \in \mathcal{F}$. Thus $\mathcal{F}$ is a normal family by Marty theorem (Theorem 7).

In the process to prove the above theorem, we proved the following theorem also.

Theorem 9 (Montel theorem)   If $\mathcal{F}$ is a family of holomorphic functions on Ω, every function in $\mathcal{F}$ does not assume two distinct complex numbers, then $\mathcal{F}$ is a normal family.

§ 5.6   Picard Great Theorem

Theorem 3 in Chapter III §3.2 is the Weierstrass theorem: If $f(z)$ is holomorphic on $D'(0, r) = D(0, r)\backslash\{0\}$, and $z = 0$ is an essential singularity of $f(z)$, then the values of $f(z)$ on $D'(0, r)$ is dense on $\mathbb{C}$.

Picard great theorem describes the value distribution of the function in the neighborhood of the essential singularity of the function more clear and deepening.

Theorem 10 (Picard great theorem)   If $f(z)$ is holomorphic on $D'(0, r)$, and $z = 0$ is an essential singularity of $f(z)$, then in each neighborhood of $z = 0$, f assumes each complex number, with one possible exception. Of course, we may replace $z = 0$ by any other point.

Picard great theorem deepens Weierstrass theorem, and is the extension of Picard little theorem.

In Chapter III §3.2 we already know that, if $f(z)$ is an entire function, the point at infinity is its pole, then $f(z)$ is a polynomial, $f(z)$ may assume any value in $\mathbb{C}$ by the fundamental theorem of algebra (Chapter II §2.4 Theorem 10). If the point at infinity is a removable singularity of $f(z)$, then $f(z)$ is a bounded entire function, and hence $f(z)$ is a constant by Liouville theorem. If the point at infinity is an essential singularity of $f(z)$, then $f(z)$ assumes any value of $\mathbb{C}$ except one value on the neighborhood of the point at infinity by Picard great Theorem. This is the Picard little theorem.

Thus Picard little theorem is a consequence of the Picard great theorem.

We use the results in the last section to prove the Picard great theorem.

Proof of Theorem 10   If Picard great theorem is not true, without loss of generality, we may suppose $f(z)$ is holomorphic on $D'(0,1)$, and f does not assume 0 and 1 on $D'(0,1)$, then we can prove that $z = 0$ is a removable singularity or a pole.

Let $f_n(z) = f(\frac{z}{n})$ $(0 < |z| < 1)$, and let $\mathcal{F} = \{f_n\}$. Every function of $\mathcal{F}$ assume $\mathbb{C}_{0,1}$, $\mathcal{F}$ is a normal family by Theorem 9. There exists a subsequence $\{f_{n_k}\}$ in $\{f_n\}$, which is normally convergent or compactly divergent. If $\{f_{n_k}\}$ is normally convergent, then $\{f_{n_k}\}$ converges uniformly on any compact subset in $D'(0,1)$, hence it is bounded. Especially it is bounded on $\{z \mid |z| = \frac{1}{2}\}$. If its upper bound is M. Thus $f(z)$ has its upper bound M on $\{z \mid |z| = \frac{1}{2n_k}\}$. By the Maximum modulus theorem, f has an upper bound M on $0 < |z| < \frac{1}{2}$. It implies that $z = 0$ is a removable singularity by Theorem 9 of Chapter II §2.3.

If $\{f_{n_k}\}$ is compact divergent, we may prove that $\frac{1}{f} \to 0$ when $z \to 0$ by the same method as we used above. That means $f \to \infty$ as $z \to 0$. Then $z = 0$ is a pole of $f(z)$.

We have proved the theorem.

The Picard great theorem and Picard little theorem are two of the most important theorems in value distribution, usually it does not appear in the undergraduate textbook, because the proof of these theorems need elliptic modulus function, it is complicated and difficult. After Picard established his theorems, there are many simplify proofs. In this Chapter we used the proof by differential geometry. 1938, L.V. Ahlfors[5] established the important Ahlfors-Schwarz lemma (Theorem 1). According to the idea of Ahlfors, in 1939, R.M. Robinson[6] proved Picard theorems by the method of differential geometry and he avoided the elliptic modulus function. After that, there are many important progress, for example, the works of H. Grauert and H. Reckziegel[7], Z. Kobayashi[8], L. Zalcman[9], D. Minda and G. Schober[10], and S.G. Krantz[11], etc. We refered these works, especially, the works of Minda and Schober, and Krantz, to write this chapter. It gave a simple proof of Picard theorem, and also let the reader start to treat the complex analysis by differential geometry. Moreover, we may use the method of differential geometry to prove some other important theorems in complex analysis, for example, Bloch theorem, Landau theorem, Schottky theorem, etc. Now we state Bloch theorem, Landau theorem and Schottky theorem as follows and omit the detail of proofs. The readers may refer the above references and L.V. Ahlfors[5,12], J.B. Conway[13]

Bloch theorem   If $f(z)$ is holomorphic on the unit disk D, $f'(0) = 1$, then $f(D)$ contains a disk with radius B, where B is a positive constant which is independence of f.

Landau theorem   If $f(z) = a_0 + a_1 z + \cdots$ $(a_1 \neq 0)$ is holomorphic on $D(0, r)$, and f does not assume 0 and 1, then $r \leq R(a_0, a_1)$ where $R(a_0, a_1)$ is a constant depending on a_0 and a_1 only.

Schottky theorem   If $f(z) = a_0 + a_1 z + \cdots$ is holomorphic on $D(0, r)$, and f does not assume 0 and 1, then for every $\theta \in (0, 1)$, there exists a constant $M(a_0, \theta)$ depending on a_0 and θ only, so that $|f(z)| \leq M(a_0, \theta)$ holds for all $|z| \leq \theta r$.

We also may prove that: Bloch theorem implies the Picard little theorem, Schottky theorem implies the Picard great theorem.

In the proof of Picard little theorem and Picard great theorem, we constructed a metric μ on $\mathbb{C}_{0,1}$. This is a key step of the proofs. This metric need satisfied the following conditions: its curvature has an negative upper bound, and there exists a positive constant M such that $\sigma \leq M\mu$. Of course, the metric (4.1) satisfied these two condition, but we may construct other metrics to satisfy these two conditions. The reader may refer the paper of R.M. Robinson[6].

EXERCISES V

1. Show that: $\Delta = \frac{\partial^2}{\partial r^2} + \frac{1}{r} \frac{\partial}{\partial r} + \frac{1}{r^2} \frac{\partial^2}{\partial \theta^2}$, where $z = re^{i\theta}$.

2. Prove Theorem 2.

3. Show that the Euclidean metric is invariant under the group of Euclidean motions.

4. If $P = (x_1, x_2, x_3)$ and $P' = (x'_1, x'_2, x'_3)$ are two points on the Riemann sphere. Show that the length of the great circular arc $\overset{\frown}{PP'}$ is

$$d(P, P') = 2\arctan \sqrt{\frac{1 - x_1 x'_1 - x_2 x'_2 - x_3 x'_3}{1 + x_1 x'_1 + x_2 x'_2 + x_3 x'_3}}.$$

If z, z' are the corresponding points of P, P' respectively by stereographic projection, then $d(P, P')$ equals to

$$d(z, z') = 2\arctan\left|\frac{z - z'}{1 + z\bar{z}'}\right|$$

by (1.5), and using this result to prove the corresponding metric is

$$ds^2 = \frac{4|dz|^2}{(1 + |z|^2)^2}.$$

5. Evaluate that the curvature of Euclidean metric is 0, the curvature of Poincaré metric is -1, and the curvature of spherical metric is $+1$.

6. Verify that the Picard little theorem is true for the function e^z.

7. What is the exception value of the function $e^z + 1$?

8. Show that: functions $\operatorname{ch} z$ and $\operatorname{sh} z$ assume all complex numbers.

9. Verify the Picard great theorem for the function $e^{1/z}$ at the neighborhood of $z = 0$.

APPENDIX   Curvature

The reader may find the knowledge about the curvature of a surface in 3-dimensional Euclidean space in any undergraduate textbook of differential geometry. Here we give a sketch of it. The reader may omit it if he know it already.

Let $D \subset \mathbb{R}^2$ be a domain, $(u, v) \in D$. If

$$\vec{r}(u, v) = \big(x(u, v), y(u, v), z(u, v)\big) \in \mathbb{R}^3$$

and x, y, z have the second order partial derivatives for u and v, $\vec{r}_u \times \vec{r}_v \neq 0$, where $\vec{r}_u, \vec{r}_v$ means the partial derivatives of $\vec{r}$ with respect to u and v respectively, then $S = \vec{r}(D)$ is a surface in $\mathbb{R}^3$, (u, v) is the parameter of S. If $(u, v) \in D$, then $\vec{r}(u, v)$ is a point P on S, $\vec{r}_u, \vec{r}_v$ forms a base of the tangent plane $T_P S$ of S at P. The **first fundamental form** of surface S is

$$I = E(du)^2 + 2F du dv + G(dv)^2 = ds^2,$$

where $E = \vec{r}_u \cdot \vec{r}_u$, $F = \vec{r}_u \cdot \vec{r}_v$, $G = \vec{r}_v \cdot \vec{r}_v$. It denotes the square of the length of the small vector $d\vec{r} = \vec{r}_u du + \vec{r}_v dv$.

If $Q : \vec{r}(u + \Delta u, v + \Delta v)$ is a point closed P, then

$$\vec{PQ} = \vec{r}(u + \Delta u, v + \Delta v) - \vec{r}(u, v)$$
$$= \vec{r}_u \Delta u + \vec{r}_v \Delta v + \frac{1}{2}\left(\vec{r}_{uu}(\Delta u)^2 + 2\vec{r}_{uv}\Delta u \Delta v + \vec{r}_{vv}(\Delta v)^2\right) + \cdots,$$

where $\vec{r}_{uu}, \vec{r}_{uv}, \vec{r}_{vv}$ denote the second derivative of $\vec{r}$ with respect to u, the second derivative of $\vec{r}$ with respect to u and v, and the second derivative of $\vec{r}$ with respect to v respectively. The direction distance from point Q to $T_P S$ is

$$\delta = \vec{PQ} \cdot \vec{n} = \frac{1}{2}\left(\vec{r}_{uu}(\Delta u)^2 + 2\vec{r}_{uv}\Delta u \Delta v + \vec{r}_{vv}(\Delta v)^2\right)\vec{n} + \cdots,$$

where $\vec{n} = \frac{\vec{r}_u \times \vec{r}_v}{|\vec{r}_u \times \vec{r}_v|}$ denotes the normal vector of surface S at point P, the terms which have not explicit expressed are the high order of Δu and Δv. The principal part of 2δ is the **second fundamental form** of surface S, it is

$$\mathrm{II} = L(\mathrm{d}u)^2 + 2M\,\mathrm{d}u\,\mathrm{d}v + N(\mathrm{d}v)^2,$$

where

$$L = \frac{\vec{r}_{uu} \cdot \vec{r}_u \times \vec{r}_v}{|\vec{r}_u \times \vec{r}_v|}, \qquad M = \frac{\vec{r}_{uv} \cdot \vec{r}_u \times \vec{r}_v}{|\vec{r}_u \times \vec{r}_v|}, \qquad N = \frac{\vec{r}_{vv} \cdot \vec{r}_u \times \vec{r}_v}{|\vec{r}_u \times \vec{r}_v|}.$$

We know that $\vec{r}_u \cdot \vec{n} = 0$ and $\vec{r}_v \cdot \vec{n} = 0$. Differentiating both sides of this two equalities, we have

$$\vec{r}_{uu} \cdot \vec{n} + \vec{r}_u \cdot \vec{n}_u = 0, \qquad \vec{r}_{uv} \cdot \vec{n} + \vec{r}_u \cdot \vec{n}_v = 0,$$
$$\vec{r}_{uv} \cdot \vec{n} + \vec{r}_v \cdot \vec{n}_u = 0, \qquad \vec{r}_{vv} \cdot \vec{n} + \vec{r}_v \cdot \vec{n}_v = 0.$$

Thus $\mathrm{II} = -\mathrm{d}\vec{r} \cdot \mathrm{d}\vec{n}$, the second fundamental form describes the curved situation of the surface at a point P.

The important facts are: I and II are geometric invariant. It means that if we use another parameter $(\bar{u}, \bar{v}) \in D$ to represent the surface S, then I and II are invariant. The fundamental theorem of curve is that if any two differentiable functions $f(x) > 0$ and $g(x)$ are given, then there exists uniquely a curve which admits f and g as the curvature and the torsion of this curve respectively if the initial conditions are given. In the theory of surface, if any six functions are given, and satisfy certain conditions (Gauss equation and Codazzi equation), then up to a motion in the space there exists uniquely a surface such that the coefficients of its first and second fundamental forms are these given six functions. This is the fundamental theorem of surface.

If $\vec{r} = \vec{r}(s)$ is a curve C on the surface S which passes the point P and its parameters is the length of the curve, then

$$\vec{T}(s) = \frac{d\vec{r}}{ds} = \vec{r}_u \frac{du}{ds} + \vec{r}_v \frac{dv}{ds}$$

is the unit tangent vector on S at P, and

$$\vec{T}'(s) = \frac{d^2\vec{r}}{ds^2} = \vec{r}_{uu}\left(\frac{du}{ds}\right)^2 + 2\vec{r}_{uv}\frac{du}{ds}\frac{dv}{ds} + \vec{r}_{vv}\left(\frac{dv}{ds}\right)^2 + \vec{R}$$

is a normal vector orthogonal $\vec{T}(s)$ where $\vec{R}$ is a tangent vector of S at P. If $\vec{N}(s)$ is the unit normal vector of curve C at P, then $T'(s) = k\vec{N}(s)$. The value

$$k_n = k\vec{N} \cdot \vec{n} = \vec{r}_{uu}\left(\frac{du}{ds}\right)^2 + 2\vec{r}_{uv}\left(\frac{du}{ds}\right)\left(\frac{dv}{ds}\right) + \vec{r}_{vv}\left(\frac{dv}{ds}\right)^2$$

is the **normal curvature** of the curve C at point P, where $\vec{n}$ is the unit normal vector of S at P. Obviously,

$$k_n = \frac{Ldu^2 + 2Mdu\,dv + Ndv^2}{Edu^2 + 2Fdu\,dv + Gdv^2},$$

it is the quotient of the second fundamental form and the first fundamental form. Of course, k_n is dependent on the directions du and dv. There are directions which make k_n extreme. We call these directions as the principal directions, and the extreme values as **principal curvatures**. In general we may define two curvatures, one is **Gaussian curvature** K, the product of two principal curvatures; another one is the **mean curvature** H, the arithmetic mean of two principal curvatures. Now we evaluate these two curvatures.

Obviously $\left(\begin{smallmatrix} E & F \\ F & G \end{smallmatrix}\right)$ is a positive definition matrix, we may prove that

(1) $\det\left(\lambda\left(\begin{smallmatrix} E & F \\ F & G \end{smallmatrix}\right) - \left(\begin{smallmatrix} L & M \\ M & N \end{smallmatrix}\right)\right) = 0$ has two real roots λ_1 and λ_2;

(2) If $\lambda_1 \neq \lambda_2$, then there exists $A = \left(\begin{smallmatrix} a_{11} & a_{12} \\ a_{21} & a_{22} \end{smallmatrix}\right)$ such that

$$A\begin{pmatrix} E & F \\ F & G \end{pmatrix}A' = \begin{pmatrix} 1 & 0 \\ 0 & 1 \end{pmatrix}, \qquad A\begin{pmatrix} L & M \\ M & N \end{pmatrix}A' = \begin{pmatrix} \lambda_1 & 0 \\ 0 & \lambda_2 \end{pmatrix}$$

hold where A' is the transpose of A.

Let $(d\bar{u}, d\bar{v}) = (du, dv)A^{-1}$, then

$$k_n = \frac{\lambda_1(d\bar{u})^2 + \lambda_2(d\bar{v})^2}{(d\bar{u})^2 + (d\bar{v})^2}.$$

Let $\bar{\lambda} = \max(\lambda_1, \lambda_2)$, $\underline{\lambda} = \min(\lambda_1, \lambda_2)$, then $\underline{\lambda} \le k_n \le \bar{\lambda}$ holds for any direction du, dv. We may prove that there exists a direction such that its normal curvatures are λ_1 and λ_2, thus λ_1 and λ_2 are principal curvatures. We have $K = \lambda_1\lambda_2$ and $H = \frac{\lambda_1+\lambda_2}{2}$ By (1), we obtain

$$K = \frac{LN - M^2}{EG - F^2}, \qquad H = \frac{1}{2} \cdot \frac{GL - 2FM + EN}{EG - F^2}.$$

The above formulas hold true again when $\lambda_1 = \lambda_2$.

In differential geometry, the curvature is the most important object to discuss, especially the Gaussian curvature. In textbooks, curvature usually means the Gaussian curvature.

Now we evaluate the Gaussian curvature when the first fundamental form, the metric of S is $ds^2 = \rho^2 du^2 + \rho^2 dv^2$. In this case $E = G = \vec{r}_u \cdot \vec{r}_u = \vec{r}_v \cdot \vec{r}_v = \rho^2$, $F = \vec{r}_u \cdot \vec{r}_v = 0$, hence

$$|\vec{r}_u \times \vec{r}_v|^2 = (\vec{r}_u \cdot \vec{r}_u)(\vec{r}_v \cdot \vec{r}_v) - (\vec{r}_u \cdot \vec{r}_v) = \rho^2,$$

it is $|\vec{r}_u \times \vec{r}_v| = \rho$, and it is easy to verify

$$LN = \frac{1}{\rho^4} \begin{vmatrix} \vec{r}_{uu} \cdot \vec{r}_{vv} & \vec{r}_{uu} \cdot \vec{r}_u & \vec{r}_{uu} \cdot \vec{r}_v \\ \vec{r}_u \cdot \vec{r}_{vv} & \rho^2 & 0 \\ \vec{r}_v \cdot \vec{r}_{vv} & 0 & \rho^2 \end{vmatrix}$$

and

$$M^2 = \frac{1}{\rho^4} \begin{vmatrix} \vec{r}_{uv} \cdot \vec{r}_{uv} & \vec{r}_{uv} \cdot \vec{r}_u & \vec{r}_{uv} \cdot \vec{r}_v \\ \vec{r}_u \cdot \vec{r}_{uv} & \rho^2 & 0 \\ \vec{r}_v \cdot \vec{r}_{uv} & 0 & \rho^2 \end{vmatrix}$$

Differentiating $\vec{r}_u \cdot \vec{r}_u = \vec{r}_v \cdot \vec{r}_v = \rho^2$, $\vec{r}_u \cdot \vec{r}_v = 0$ with respect to u and v respectively, we have

$$\vec{r}_{uu} \cdot \vec{r}_u = \rho\rho_u, \qquad \vec{r}_{uv} \cdot \vec{r}_u = \rho\rho_v, \qquad \vec{r}_{uv} \cdot \vec{r}_v = \rho\rho_v,$$

$$\vec{r}_{vv} \cdot \vec{r}_v = \rho\rho_v, \qquad \vec{r}_{uu} \cdot \vec{r}_v = -\rho\rho_v, \qquad \vec{r}_u \cdot \vec{r}_{uv} = -\rho\rho_u$$

and

$$\frac{\partial}{\partial v}(\vec{r}_{uu} \cdot \vec{r}_v) = \vec{r}_{uuv} \cdot \vec{r}_v + \vec{r}_{uu} \cdot \vec{r}_{vv} = -\rho_v^2 - \rho\rho_{vv},$$

$$\frac{\partial}{\partial u}(\vec{r}_{uv} \cdot \vec{r}_v) = \vec{r}_{uuv} \cdot \vec{r}_v + \vec{r}_{uv} \cdot \vec{r}_{uv} = \rho_u^2 + \rho\rho_{uu}.$$

Hence

$$\vec{r}_{uu} \cdot \vec{r}_{vv} - \vec{r}_{uv} \cdot \vec{r}_{uv} = -(\rho_u^2 + \rho_v^2) - \rho\Delta\rho.$$

Substituting these results into the determinants which express LN and M^2, we obtain

$$
\begin{aligned}
LN - M^2 = &\frac{1}{\rho^4}
\begin{vmatrix}
\vec{r}_{uu} \cdot \vec{r}_{vv} & \rho\rho_u & -\rho\rho_v \\
-\rho\,\rho_u & \rho^2 & 0 \\
\rho\rho_v & 0 & \rho^2
\end{vmatrix}
- \frac{1}{\rho^4}
\begin{vmatrix}
\vec{r}_{uv} \cdot \vec{r}_{uv} & \rho\rho_v & -\rho\rho_u \\
\rho\rho_v & \rho^2 & 0 \\
\rho\rho_u & 0 & \rho^2
\end{vmatrix} \\
= &\vec{r}_{uu} \cdot \vec{r}_{vv} + \rho_u^2 + \rho_v^2 - \vec{r}_{uv} \cdot \vec{r}_{uv} + \rho_u^2 + \rho_v^2 \\
= &\rho_u^2 + \rho_v^2 - \rho\Delta\rho.
\end{aligned}
$$

Thus the Gaussian curvature is

$$\frac{\rho_u^2 + \rho_v^2}{\rho^4} - \frac{\Delta\rho}{\rho^3} = -\frac{1}{\rho^2}\Delta\log\rho$$

since $EG - F^2 = \rho^4$.

CHAPTER VI
ELEMENTARY FACTS ON SEVERAL
COMPLEX VARIABLES

§ 6.1 Introduction

In the last chapter of this book, we will study a few results in several complex variables, in purpose to describe the essential difference between the theory of function of one complex variable and several complex variables. By these observation, we may understand one complex variable more deep.

Just like the other theories in mathematics, to extend the theory from one dimension to high dimension, some parts are easy to parallel extend, usually this part is not difficult, but some parts happen in high dimension only. Of course, the second parts are more important.

Just like the theory of one complex variable, the theory of several complex variables has a long history. Around the beginning of 20th century, two important theorems in several complex variables were discovered, it opened a new page in several complex variables. These two theorems are Poincaré theorem and Hartogs theorem. Let $\mathbb{C}^n$ be n-dimensional complex Euclidean space. If $\Omega \subset \mathbb{C}^n$ is a domain, $z = (z_1, z_2, \cdots, z_n) \in \Omega$, and $g(z) : \Omega \to \mathbb{C}$, g is **holomorphic** if for each $j = 1, \cdots, n$ and each fixed $z_1, \cdots, z_{j-1}, z_{j+1}, \cdots, z_n$, the function $g(z_1, \cdots, z_{j-1}, \zeta, z_{j+1}, \cdots, z_n)$ is a holomorphic function of ζ in the classical one-variable sense on the set $\{\zeta \in \mathbb{C} \mid (z_1, \cdots, z_{j-1}, \zeta, z_{j+1}, \cdots, z_n) \in \Omega\}$. In other words, f is holomorphic in each variable separately. The mapping $f(z) = (f_1(z), f_2(z), \cdots, f_n(z))$ is holomorphic on a domain (connected open set) $\Omega \subseteq \mathbb{C}^n$ if each $f_j(z)$ $(j = 1, 2, \cdots, n)$ is a holomorphic function on Ω. Moreover, the holomorphic mapping is **biholomorphic** if it is one to one, onto and f^{-1} is holomorphic (the last condition is redundant, such a mapping automatically has a holomorphic inverse, the proof is complicated, we omit it). Poincaré theorem is as follows. In $\mathbb{C}^n$ $(n \geq 2)$, the unit ball $B(0, 1)$ and unit polydisc $D^n(0, 1)$ are not biholomorphic, where $B(0, 1) = \{z \in \mathbb{C}^n \mid |z_1|^2 + \cdots + |z_n|^2 < 1\}$, $D^n(0, 1) = \{z \in \mathbb{C}^n \mid |z_1| < 1, \cdots, |z_n| < 1\}$. That means there does not exist a biholomorphic mapping of $B(0,1)$ onto $D^n(0,1)$. Poincaré Theorem tells us that in high dimension, the Riemann mapping theorem in Chapter IV never holds.

The Hartogs theorem is as follows. In $\mathbb{C}^n$ ($n \geq 2$), there exists such a kind of domain, if a function is holomorphic on this domain, then the function is holomorphic on a larger domain which contains this domain. That means, the function can be analytically continued to a larger domain. This phenomenon does not happen in one complex variable. A basic question raising from Hartogs theorem is that what kind domains we will study in function theory in several complex variables.

After more than two hundred years development, the theory of function of one complex variable is mature, but the understanding about the several complex variables is just started.

In this chapter, we state some results of several complex variables which are easy extended from one complex variable at first, then we prove the Poincaré theorem and Hartog theorem as a brick to knock the door of several complex variables. For simplify, we only discuss the Euclidean space of two complex variables, $\mathbb{C}^2 = \mathbb{C} \times \mathbb{C}$, and to prove Hartogs theorem in a special case only.

We state the following theorems and omit the detail of the proofs, because the method of proofs is similar with one complex variable. The readers can prove it if he refers the proof of the corresponding theorem in one complex variable.

Theorem 1 (Cauchy integral formula) Let $D^2(w,r) = \{z = (z_1, z_2) \in \mathbb{C}^2 \mid |z_1 - w_1| < r, |z_2 - w_2| < r\}$ be a bidisk centered at w, with radius r, where $w = (w_1, w_2) \in \mathbb{C}^2$, $r > 0$. If $f(z)$ is holomorphic on $\overline{D^2(w,r)}$, then

$$f(z) = \frac{1}{(2\pi i)^2} \int_{\partial D(w_1,r)} \int_{\partial D(w_2,r)} \frac{f(\zeta_1, \zeta_2)}{(\zeta_1 - z_1)(\zeta_2 - z_2)} \, d\zeta_1 d\zeta_2 \qquad (1.1)$$

holds for any $z \in D^2(w,r)$.

We obtain (1.1) by using Cauchy integral formula of one complex variable twice.

Just like in one complex variable, we know that the partial derivatives of any order of $f(z)$ exist, and

$$
\left(\frac{\partial}{\partial z_1}\right)^j \left(\frac{\partial}{\partial z_2}\right)^k f(z)
$$
$$
= \frac{j!\,k!}{(2\pi i)^2} \int_{\partial D(w_1,r)} \int_{\partial D(w_2,r)} \frac{f(\zeta_1,\zeta_2)}{(\zeta_1 - z_1)^{j+1}(\zeta_2 - z_2)^{k+1}} \, d\zeta_1 \, d\zeta_2
$$

holds for any non-negative integers j and k, we have Cauchy inequality

$$
\left|\left(\frac{\partial}{\partial z_1}\right)^j \left(\frac{\partial}{\partial z_2}\right)^k f(z)\right| \le \frac{Mj!\,k!}{r^{j+k}},
$$

where

$$
M = \sup_{|\zeta_1 - z_1| = r,\, |\zeta_2 - z_2| = r} |f(z)|, \qquad \zeta = (\zeta_1, \zeta_2).
$$

Just like in one complex variable, if $f(z)$ is holomorphic in a neighborhood of $\overline{D^2(w,r)}$, then $f(z)$ can be expanded as a Taylor series

$$
f(z) = \sum_{j,k=0}^{\infty} a_{jk}(z_1 - w_1)^j (z_2 - w_2)^k
$$

and it converges uniformly and absolutely on $\overline{D^2(w,r)}$, and

$$
a_{jk} = \frac{1}{j!\,k!}\left(\frac{\partial}{\partial z_1}\right)^j \left(\frac{\partial}{\partial z_2}\right)^k f(w).
$$

Just like in one complex variable, we may prove the following theorems.

Theorem 2   If $f(z)$ is holomorphic on domain $U \subseteq \mathbb{C}^n$, and $f(z)$ equals zero on an open set in U, then $f(z)$ identically equals to zero on U.

Theorem 3 (Maximum modulus principle)   If U is a bounded domain in $\mathbb{C}^2$, $f(z)$ is holomorphic on U, $M = \sup\limits_{\zeta \in \partial U} \lim\limits_{\substack{z \to \zeta \\ z \in U}} |f(z)|$, then $|f(z)| < M$ for $z \in U$, unless f is a constant.

Similar with Theorem 1 of Chapter III §3.1 (Weierstrass theorem), we have the following theorems.

Theorem 4 (Weierstrass theorem)   If Ω is a domain in $\mathbb{C}^2$, $\{f_n\}$ is a sequence of holomorphic functions on Ω, and it converges uniformly on any compact subset in Ω, then $f = \lim\limits_{n \to \infty} f_n$ is holomorphic on Ω, and $\{(\frac{\partial}{\partial z_1})^j (\frac{\partial}{\partial z_2})^k f_n\}$ converges uniformly to $(\frac{\partial}{\partial z_1})^j (\frac{\partial}{\partial z_2})^k f$ on any compact subset in Ω.

Similar with Theorem 3 of Chapter VI §4.2 (Montel theorem), we have the following theorem.

Theorem 5 (Montel theorem) If $\mathcal{F} = \{f_n\}$ is a family of holomorphic functions on $\Omega \subseteq \mathbb{C}^2$, and there exists a positive constant $M > 0$, such that $|f(z)| \le M$ holds for all $z \in \Omega$, $f \in \mathcal{F}$, then for any sequence $\{f_n\}$ in $\mathcal{F}$, we may find a subsequence which converges uniformly on any compact subset in Ω. Thus $\mathcal{F}$ is a normal family.

§ 6.2   Cartan Theorem

We already decided the group of holomorphic automorphisms of the unit disc in Chapter II §2.5. Now we want to decide the group of holomorphic automorphisms of the unit ball and the unit bidisk. Then we use it to prove the Poincaré theorem.

If we recall how to decide the group of holomorphic automorphisms of the unit disk, we may find that its main idea is to use the Schwarz lemma. In the case of several complex variables, we need to use the extension of Schwarz lemma, it is two theorems of Cartan.

Theorem 6 (Cartan theorem) If $U \subseteq \mathbb{C}^2$ is a bounded domain, $P \in U$, and $f = (f_1, f_2)$ is a holomorphic mapping of U into U, and $f(P) = P$, $J_f(P) = I$, then $f(z) \equiv z$, where $J_f(z)$ is the Jacobi matrix of f at z,

$$\begin{pmatrix} \dfrac{\partial f_1}{\partial z_1} & \dfrac{\partial f_1}{\partial z_2} \\ \dfrac{\partial f_2}{\partial z_1} & \dfrac{\partial f_2}{\partial z_2} \end{pmatrix},$$

I is the identity matrix, i.e.,

$$I = \begin{pmatrix} 1 & 0 \\ 0 & 1 \end{pmatrix}.$$

Proof Without loss of generality, we may assume $P = 0$. If the theorem does not hold true, then we may expand $f(z)$ at $z = 0$ as a Taylor series

$$f(z_1, z_2) = z + A_m(z) + \cdots$$

where $A_m(z) = \left(A_m^{(1)}(z), A_m^{(2)}(z) \right)$ is the first non-zero term, $A_m^{(1)}(z), A_m^{(2)}(z)$ are homogeneous polynomials of z_1 and z_2 of order m $(m \ge 2)$.

Let $f^1 = f,\ f^2 = f \circ f, \cdots, f^j = f^{j-1} \circ f\ (j \geq 2)$, then

$$
\begin{aligned}
f^1(z) &= z + A_m(z) + \cdots, \\
f^2(z) &= z + A_m(f(z)) + \cdots \\
&= z + A_m(z) + A_m(z) + \cdots \\
&= z + 2A_m(z) + \cdots,
\end{aligned}
\tag{2.1}
$$

$$\cdots$$

$$f^j(z) = z + j A_m(z) + \cdots.$$

Since U is a bounded domain and f maps U into U, $\{f^j\}$ forms a normal family by Theorem 5 (Montel Theorem). That means: there exists a subsequence $\{f^{j_l}\}$, so that $f^{j_l} \to F$ when $j_l \to \infty$. By Theorem 4 (Weierstrass theorem), the value of the derivatives of order m of f^{j_l} at $z = 0$ converges to the value of the derivative of order m of F at $z = 0$. From (2.1), the value of the derivatives of order m of f^{j_l} at $z = 0$ approaches to ∞ when $j_l \to \infty$. But the value of the derivative of order m of F at $z = 0$ is impossible equal to ∞. We obtain a contradiction. Thus $A_m(z) = 0$ on U. That is $f(z) \equiv z$.

This Cartan theorem (Theorem 6), in the case of one complex variable, U becomes the unit disk D, and Cartan theorem becomes that if holomorphic function $f(z)$ maps D into D, $f(0) = 0$, $f'(0) = 1$, then $f(z) \equiv z$. That is, the equality case of the Schwarz lemma.

Next we state and prove another Cartan theorem.

If $U \subseteq \mathbb{C}^2$ is a domain, U is a **circular domain** if $(\mu z_1, \mu z_2) \in U$ whenever $(z_1, z_2) \in U$ and μ is any complex number with $|\mu| < 1$.

Theorem 7 (Cartan theorem)   If $U \subseteq \mathbb{C}^2$ is a bounded circular domain, f is a biholomorphic mapping of U onto U, and $f(0) = 0$, then f is a linear mapping, that is $f(z) = zA$ where A is a constant matrix.

Proof   Let $\theta \in [0, 2\pi]$, $\rho_\theta(z_1, z_2) = (e^{i\theta} z_1, e^{i\theta} z_2)$, consider the mapping $g = \rho_{-\theta} \circ f^{-1} \circ \rho_\theta \circ f$, then

$$
J_g(0) = \begin{pmatrix} e^{-i\theta} & 0 \\ 0 & e^{-i\theta} \end{pmatrix} J_f^{-1}(0) \begin{pmatrix} e^{i\theta} & 0 \\ 0 & e^{i\theta} \end{pmatrix} J_f(0) = I,
$$

and $g(z)$ maps U into U, $g(0) = 0$. Thus $g(z) \equiv z$ by Theorem 6, we have

$$f \circ \rho_\theta = \rho_\theta \circ f. \tag{2.2}$$

Expand f at a neighborhood of $z = 0$ as power series,

$$f(z) = \sum_{j,k=0}^{\infty} a_{jk} z_1^j z_2^k = \left(\sum_{j,k=0}^{\infty} a_{jk}^{(1)} z_1^j z_2^k, \ \sum_{j,k=0}^{\infty} a_{jk}^{(2)} z_1^j z_2^k \right),$$

we have

$$\rho_\theta \circ f = (e^{i\theta} f_1, e^{i\theta} f_2) = \left(\sum_{j,k=0}^{\infty} a_{jk}^{(1)} e^{i\theta} z_1^j z_2^k, \ \sum_{j,k=0}^{\infty} a_{jk}^{(2)} e^{i\theta} z_1^j z_2^k \right)$$

and

$$f \circ \rho_\theta = \left(\sum_{j,k=0}^{\infty} a_{jk}^{(1)} (e^{i\theta} z_1)^j (e^{i\theta} z_2)^k, \ \sum_{j,k=0}^{\infty} a_{jk}^{(2)} (e^{i\theta} z_1)^j (e^{i\theta} z_2)^k \right)$$

$$= \left(\sum_{j,k=0}^{\infty} a_{jk}^{(1)} e^{i(j+k)\theta} z_1^j z_2^k, \ \sum_{j,k=0}^{\infty} a_{jk}^{(2)} e^{i(j+k)\theta} z_1^j z_2^k \right).$$

Using (2.2) and comparing the corresponding coefficients, we have $a_{jk} = 0$ except $j + k = 1$. Thus $f(z)$ is linear at a neighborhood of $z = 0$. Hence $f(z)$ is linear on U by Theorem 2.

This Cartan theorem (Theorem 7), in the case of one complex variable, U becomes the unit disk D, and Cartan theorem becomes that if holomorphic function $f(z)$ maps D onto D, and $f(0) = 0$, then $f(z) = e^{i\theta} z$.

We may use these two Cartan theorems to decide the groups of holomorphic automorphisms of the unit ball and the unit bidisk in $\mathbb{C}^2$.

§ 6.3 Groups of Holomorphic Automorphisms of Unit Ball and Unit Bidisk

Let U be a domain in $\mathbb{C}^2$. If there exists a biholomorphic mapping $f(z)$ of U onto itself, then $f(z)$ is a holomorphic automorphism of U, or a biholomorphic automorphism of U. All holomorphic automorphisms of U forms a group, it is the group of holomorphic automorphisms of U, and denote it by Aut (U).

Theorem 8   Aut $(D^2(0,1))$ is composed by all biholomorphic mappings

$$w = \left(e^{i\theta_1} \frac{z_1 - a_1}{1 - \bar{a}_1 z_1}, \ e^{i\theta_2} \frac{z_2 - a_2}{1 - \bar{a}_2 z_2} \right) \tag{3.1}$$

and

$$w = \left(e^{i\theta_2}\frac{z_2 - a_1}{1 - \bar{a}_1 z_2}, \ e^{i\theta_1}\frac{z_1 - a_2}{1 - \bar{a}_2 z_1}\right), \tag{3.2}$$

where $z = (z_1, z_2) \in D^2(0, 1)$, $a_1, a_2 \in D(0, 1)$ and $\theta_1, \theta_2 \in [0, 2\pi]$.

Proof   Let $\varphi(z) \in \text{Aut}\,(D^2(0, 1))$, and $\varphi(0) = \alpha = (\alpha_1, \alpha_2)$. Let $\psi(z) = (\frac{z_1 - \alpha_1}{1 - \bar{\alpha}_1 z_1}, \frac{z_2 - \alpha_2}{1 - \bar{\alpha}_2 z_2})$, then $g = \psi \circ \varphi \in \text{Aut}\,(D^2(0, 1))$, and $g(0) = 0$. By Theorem 7 (Cartan theorem), we have

$$g(z) = zA = (z_1, z_2)\begin{pmatrix} a_{11} & a_{12} \\ a_{21} & a_{22} \end{pmatrix} = (a_{11}z_1 + a_{21}z_2, \ a_{12}z_1 + a_{22}z_2).$$

Hence

$$|a_{11}z_1 + a_{21}z_2| < 1, \qquad |a_{12}z_1 + a_{22}z_2| < 1$$

holds for any $(z_1, z_2) \in D^2(0, 1)$ since $g \in \text{Aut}\,(D^2(0, 1))$. It implies $|a_{ij}| < 1$ $(i, j = 1, 2)$. Taking

$$z^{1,k} = \left(\left(1 - \frac{1}{k}\right)\frac{\bar{a}_{11}}{|a_{11}|}, \left(1 - \frac{1}{k}\right)\frac{\bar{a}_{21}}{|a_{21}|}\right) \in D^2(0, 1),$$

$$z^{2,k} = \left(\left(1 - \frac{1}{k}\right)\frac{\bar{a}_{12}}{|a_{12}|}, \left(1 - \frac{1}{k}\right)\frac{\bar{a}_{22}}{|a_{22}|}\right) \in D^2(0, 1),$$

then

$$g(z^{1,k}) = \left(\left(1 - \frac{1}{k}\right)(|a_{11}| + |a_{21}|), *\right) \in D^2(0, 1),$$

$$g(z^{2,k}) = \left(*, \left(1 - \frac{1}{k}\right)(|a_{12}| + |a_{22}|)\right) \in D^2(0, 1).$$

Hence

$$\left(1 - \frac{1}{k}\right)(|a_{11}| + |a_{21}|) < 1, \qquad \left(1 - \frac{1}{k}\right)(|a_{12}| + |a_{22}|) < 1.$$

Letting $k \to \infty$, we have

$$|a_{11}| + |a_{21}| \leq 1, \qquad |a_{12}| + |a_{22}| \leq 1. \tag{3.3}$$

On the other hand, we note that

$$\left(1 - \frac{1}{k}, 0\right) \in D^2(0, 1), \qquad \left(0, 1 - \frac{1}{k}\right) \in D^2(0, 1),$$

and

$$g\left(1 - \frac{1}{k}, 0\right) = \left(\left(1 - \frac{1}{k}\right)a_{11}, \left(1 - \frac{1}{k}\right)a_{12}\right) \in D^2(0, 1),$$
$$g\left(0, 1 - \frac{1}{k}\right) = \left(\left(1 - \frac{1}{k}\right)a_{21}, \left(1 - \frac{1}{k}\right)a_{22}\right) \in D^2(0, 1).$$

Since $(1 - \frac{1}{k}, 0)$ and $(0, 1 - \frac{1}{k})$ approach to $\partial D^2(0, 1)$ when $k \to \infty$, we have

$$(a_{11}, a_{12}) \in \partial D^2(0, 1), \qquad (a_{21}, a_{22}) \in \partial D^2(0, 1),$$

and hence

$$\max\{|a_{11}|, |a_{12}|\} = 1, \qquad \max\{|a_{21}|, |a_{22}|\} = 1. \tag{3.4}$$

The formulas (3.3) and (3.4) hold simultaneous only in the following two cases:
   (1) $|a_{11}| = 1$, $a_{12} = 0$, $a_{21} = 0$, $|a_{22}| = 1$,
   (2) $|a_{12}| = 1$, $a_{11} = 0$, $a_{22} = 0$, $|a_{21}| = 1$,
Thus A equals to
   (1) $A = \begin{pmatrix} e^{i\theta_1} & 0 \\ 0 & e^{i\theta_2} \end{pmatrix}$;
or
   (2) $A = \begin{pmatrix} 0 & e^{i\theta_1} \\ e^{i\theta_2} & 0 \end{pmatrix}$.
This means
   (1) $\psi \circ \varphi(z) = z \begin{pmatrix} e^{i\theta_1} & 0 \\ 0 & e^{i\theta_2} \end{pmatrix} = \left(z_1 e^{i\theta_1}, z_2 e^{i\theta_2}\right),$
or
   (2) $\psi \circ \varphi(z) = z \begin{pmatrix} 0 & e^{i\theta_1} \\ e^{i\theta_2} & 0 \end{pmatrix} = \left(z_2 e^{i\theta_2}, z_1 e^{i\theta_1}\right).$
If $\varphi = (\varphi_1, \varphi_2)$, then (1) becomes

$$\left(\frac{\varphi_1 - \alpha_1}{1 - \overline{\alpha}_1 \varphi_1}, \frac{\varphi_2 - \alpha_2}{1 - \overline{\alpha}_2 \varphi_2}\right) = \left(z_1 e^{i\theta_1}, z_2 e^{i\theta_2}\right),$$

that is

$$\frac{\varphi_1 - \alpha_1}{1 - \overline{\alpha}_1 \varphi_1} = z_1 e^{i\theta_1}, \qquad \frac{\varphi_2 - \alpha_2}{1 - \overline{\alpha}_2 \varphi_2} = z_2 e^{i\theta_2}.$$

Solving φ_1, φ_2 from these two equations, we obtain

$$\varphi = (\varphi_1, \varphi_2) = \left(e^{i\theta_1} \frac{\alpha_1 e^{-i\theta_1} + z_1}{1 + \overline{\alpha}_1 e^{i\theta_1} z_1}, e^{i\theta_2} \frac{\alpha_2 e^{-i\theta_2} + z_2}{1 + \overline{\alpha}_2 e^{i\theta_2} z_2}\right),$$

which is (3.1) if we let $\alpha_1 e^{-i\theta_1} = -a_1$, $\alpha_2 e^{-i\theta_2} = -a_2$. Similarly if we consider the case (2), we obtain (3.2).

We have decided the group of holomorphic automorphisms of $D^2(0,1)$. Next we will decide the group of holomorphic automorphisms of $B(0,1)$. Let

$$\varphi_a(z_1, z_2) = \left(\frac{z_1 - a}{1 - \bar{a}z_1}, \frac{(1 - |a|^2)^{1/2}z_2}{1 - \bar{a}z_1} \right), \tag{3.5}$$

where $a \in \mathbb{C}$, $|a| < 1$, then

$$\left| \frac{z_1 - a}{1 - \bar{a}z_1} \right|^2 + \left| \frac{(1 - |a|^2)^{1/2}z_2}{1 - \bar{a}z_1} \right|^2$$
$$= \frac{|z_1|^2 - 2\mathrm{Re}\,(\bar{a}z_1) + |a|^2 + (1 - |a|^2)|z_2|^2}{|1 - \bar{a}z_1|^2}$$
$$< \frac{1 - |a|^2 - 2\mathrm{Re}\,(\bar{a}z_1) + |a|^2 + |a|^2|z_1|^2}{|1 - \bar{a}z_1|^2} = 1$$

when $(z_1, z_2) \in B(0,1)$. Thus $\varphi_a(z_1, z_2) \in \mathrm{Aut}\,(B(0,1))$. Obviously $(\varphi_a)^{-1} = \varphi_{-a}$. A 2×2 matrix U is a unitary matrix if $U\overline{U}' = I$, where $\overline{U}'$ is the conjugate transpose of U. The mapping $w = zU$ is an unitary rotation, and denoted it by $w = U(z)$.

Theorem 9   If $g(z) \in \mathrm{Aut}\,(B(0,1))$ and $g(0) = 0$, then g is an unitary rotation, $g(z) = zA$, where A is an unitary matrix.

Proof $B(0,1)$ is a circular domain, we have $g(z) = zA$ by Theorem 7 (Cartan theorem), and g maps an unit vector to an unit vector when $g(z) \in \mathrm{Aut}\,(B(0,1))$.

If

$$A = \begin{pmatrix} a_{11}, & a_{12} \\ a_{21}, & a_{22} \end{pmatrix},$$

(α, β) is an unit vector, then

$$(\alpha, \beta) \begin{pmatrix} a_{11}, & a_{12} \\ a_{21}, & a_{22} \end{pmatrix} = (\gamma, \delta)$$

is a unit vector, and

$$\gamma = a_{11}\alpha + a_{21}\beta, \qquad \delta = a_{12}\alpha + a_{22}\beta.$$

Hence

$$|a_{11}\alpha + a_{21}\beta|^2 + |a_{12}\alpha + a_{22}\beta|^2 = 1. \tag{3.6}$$

Let $\alpha = 1$, $\beta = 0$, and then $\alpha = 0$, $\beta = 1$, we obtain

$$|a_{11}|^2 + |a_{12}|^2 = 1, \qquad |a_{21}|^2 + |a_{22}|^2 = 1. \tag{3.7}$$

Substituting (3.7) into (3.6), we have

$$\mathrm{Re}\left((a_{11}\bar{a}_{21} + a_{12}\bar{a}_{22})\alpha\bar{\beta}\right) = 0.$$

Let $\alpha = \frac{1}{\sqrt{2}}$, $\beta = \frac{1}{\sqrt{2}}$, and then $\alpha = \frac{i}{\sqrt{2}}$, $\beta = \frac{1}{\sqrt{2}}$, we obtain

$$\mathrm{Re}\,(a_{11}\bar{a}_{21} + a_{12}\bar{a}_{22}) = 0, \qquad \mathrm{Im}\,(a_{11}\bar{a}_{21} + a_{12}\bar{a}_{22}) = 0.$$

Hence

$$a_{11}\bar{a}_{21} + a_{12}\bar{a}_{22} = 0. \tag{3.8}$$

(3.7), (3.8) mean A is an unitary matrix.

Theorem 10 Every element of Aut $(B(0,1))$ can be expressed as a composition of φ_a and at most two unitary rotations.

Proof If $f \in \mathrm{Aut}\,(B(0,1))$, $f(0) = \alpha$, then there exists an unitary matrix U such that $\alpha U = (|\alpha|, 0)$. Let $g(z) = \varphi_{|a|} \circ U \circ f(z)$, where $\varphi_{|a|}$ is defined by (3.5), which maps $(|\alpha|, 0)$ to 0, then $g(z) \in \mathrm{Aut}\,(B(0,1))$, and

$$g(0) = \varphi_{|a|} \circ U \circ f(0) = \varphi_{|a|} \circ U \circ \alpha = 0.$$

By Theorem 9, we have

$$g(z) = zV = V(z),$$

where V is an unitary matrix. Thus

$$f(z) = U^{-1} \circ \varphi_{-|a|} \circ V(z).$$

We have proved Theorem 10.

§ 6.4 Poincaré Theorem

Now we prove the important Poincaré theorem.

Theorem 11 (Poincaré theorem) There is no biholomorphic mapping of the unit bidisc $D^2(0,1)$ onto the unit ball $B(0,1)$.

Proof If there exists a biholomorphic mapping φ of $D^2(0,1)$ onto $B(0,1)$, and $\varphi(0) = \alpha$, then $\Phi = \varphi_\alpha \circ \varphi$ is a biholomorphic mapping of $D^2(0,1)$ onto

$B(0,1)$ with $\Phi(0) = \varphi_\alpha \circ \varphi(0) = 0$, where $\varphi_\alpha \in \text{Aut}\,(B(0,1))$, which maps α to 0. If $h \in \text{Aut}\,(D^2(0,1))$, then

$$h \longrightarrow \Phi \circ h \circ \Phi^{-1} \in \text{Aut}\,(B(0,1)) \qquad (4.1)$$

establishs the isomorphism between these two groups. If $(\text{Aut}\,(D^2(0,1)))_0$ and $(\text{Aut}\,(B(0,1)))_0$ denote the branches of $\text{Aut}\,(D^2(0,1))$ and $\text{Aut}\,(B(0,1))$ respectively which contain the unit element, (4.1) establishs the isomorphism between $(\text{Aut}\,(D^2(0,1)))_0$ and $(\text{Aut}\,B(0,1))_0$. Let $\text{Aut}_0(D^2(0,1))$ and $\text{Aut}_0(B(0,1))$ denote the subgroups of $(\text{Aut}\,(D^2(0,1)))_0$ and $(\text{Aut}\,(B(0,1)))_0$ respectively which leave the origin invariant, (4.1) establish the isomorphism between these two subgroups.

By Theorem 8, all biholomorphic mappings

$$w = (e^{i\,\theta_1}z_1, e^{i\,\theta_2}z_2) = (z_1, z_2)\begin{pmatrix} e^{i\,\theta_1} & 0 \\ 0 & e^{i\,\theta_2} \end{pmatrix}$$

$(\theta_1, \theta_2$ are real numbers) form the subgroup $\text{Aut}_0(D^2(0,1))$. That means, $\text{Aut}_0(D^2(0,1))$ is isomorphic to the group $\{\begin{pmatrix} e^{i\,\theta_1} & 0 \\ 0 & e^{i\,\theta_2} \end{pmatrix}\}$.

By Theorem 10, all biholomorphic mappings

$$w = zVU^{-1}$$

form $\text{Aut}_0(B(0,1))$, where U, V are unitary matrices. The inverse of the unitary matrix and the product of two unitary matrices are unitary, thus all biholomorphic mappings $w = zX$ form $\text{Aut}_0(B(0,1))$ where X is an unitary matrix. That means, $\text{Aut}_0(B(0,1))$ is ismorphic to the unitary group.

If there exists biholomorphic mapping of $D^2(0,1)$ to $B(0,1)$, (4.1) establishes a group isomorphism between $\text{Aut}_0(D^2(0,1))$ and $\text{Aut}_0(B(0,1))$. It means: the group $\{\begin{pmatrix} e^{i\,\theta_1} & 0 \\ 0 & e^{i\,\theta_2} \end{pmatrix}\}$ is isomorphic to the unitary group. It is impossible, because $\{\begin{pmatrix} e^{i\,\theta_1} & 0 \\ 0 & e^{i\,\theta_2} \end{pmatrix}\}$ is an Abelian group, and the unitary group is not. Thus there is no biholomorphic mapping Φ of $D^2(0,1)$ onto $B(0,1)$. We have proved Poincaré theorem.

The Riemann mapping theorem in Chapter IV told us: If the boundary points of a simply connected domain Ω is more than one point, then there exists a holomorphic univalent function of Ω on to the unit disk. That is, the topological equivalent of two domains implies the holomorphic equivalent of two domain. Poincaré theorem told us: it is not holds true in $\mathbb{C}^n$ when $n \geq 2$.

Thus we need to consider the classification of the domains in $\mathbb{C}^n$ ($n \geq 2$), when two domains are holomorphic equivalent if they are topological equivalent. We are far from being to solve this problem. We already know that, the probability of two topological equivalent domains being holomorphic equivalent equals to zero. That is, nearly any two topological equivalent domains are not holomorphic equivalent.

These observation tells us that the Riemann mapping theorem is a very deep theorem. This kind theorem holds true only in one dimension. Starting from this important theorem, we obtained a series of deep theorems on one dimension.

§ 6.5   Hartogs Theorem

In one complex variable, if Ω is a domain in $\mathbb{C}$, $a \in \mathbb{C}\backslash\Omega$, then there exists a holomorphic function f on Ω, which can not be analytically continued to the point a. It is easy to do it, for example, $f(z) = \frac{1}{z-a}$. In several complex variables, it never holds. This phenomenon is called the Hartogs phenomenon. The Hartogs theorem can state as follows.

Theorem 12 (Hartogs theorem)   Suppose $\Omega \subseteq \mathbb{C}^n$ ($n \geq 2$) is a domain, K is a compact subset in Ω, and $\Omega\backslash K$ is connected. If f is a holomorphic function on $\Omega\backslash K$, then there exists a holomorphic function F on Ω, and F equals f on $\Omega\backslash K$.

That means, if a function is holomorphic on $\Omega\backslash K$, then it can be analytically continued to Ω.

Here we do not give the proof of the theorem in general, but we give a proof of the Hartogs theorem in some special domains.

If R is a domain in $\mathbb{C}^2$, R is a Reinhardt domain provided $(e^{i\theta_1}z_1, e^{i\theta_2}z_2) \in R$ for any two real numbers θ_1, θ_2 when $z = (z_1, z_2) \in R$.

Theorem 13   If R is a Reinhardt domain in $\mathbb{C}^2$, $f(z)$ is a holomorphic function on R, then $f(z)$ can be expanded as Laurent series

$$\sum_{j,k=-\infty}^{\infty} a_{jk} z_1^j z_2^k. \tag{5.1}$$

The series converges uniformly to f on any compact subset in R. The expansion is unique.

Proof   We prove the uniqueness at first. Fix $w = (w_1, w_2) \in R$, and $w_1 \neq 0$, $w_2 \neq 0$. Set $z_1 = w_1 e^{i\theta_1}$, $z_2 = w_2 e^{i\theta_2}$, $-\pi \leq \theta_1 \leq \pi$, $-\pi \leq \theta_2 \leq \pi$. It forms a compact subset in R. Since (5.1) converges uniformly on any compact subset in R, we have

$$a_{jk} = \frac{w_1^{-j} w_2^{-k}}{(2\pi)^2} \int_{-\pi}^{\pi} \int_{-\pi}^{\pi} f(w_1 e^{i\theta_1}, w_2 e^{i\theta_2}) e^{-i(j\theta_1 + k\theta_2)} \, d\theta_1 d\theta_2$$

for all j, k. That means, all $a_{j,k}$ are uniquely decided by f.

Next we prove the existence of (5.1).

If $f(z)$ is holomorphic on $\Omega = \{z \in \mathbb{C}^2 \mid r_1 < |z_1| < R_1, \, r_2 < |z_2| < R_2\}$, then we use Laurent expansion for one complex variable twice, we obtain the Laurent expansion

$$\sum_{j,k=-\infty}^{\infty} b_{jk} z_1^j z_2^k,$$

it converges uniformly on any compact subset in Ω.

If $w = (w_1, w_2) \in R$,

$$\Omega(w, \varepsilon) = \{z \in \mathbb{C}^2 \mid |w_1| - \varepsilon < |z_1| < |w_1| + \varepsilon, \, |w_2| - \varepsilon < |z_2| < |w_2| + \varepsilon\} \subseteq R$$

when ε is sufficiently small, we have the Laurent expansion

$$f(z) = \sum_{j,k=-\infty}^{\infty} a_{jk}(w) z_1^j z_2^k, \qquad z \in \Omega(w, \varepsilon)$$

on $\Omega(w, \varepsilon)$, and it converges to f at a neighborhood of w.

If $w' \in \Omega(w, \varepsilon)$, and the Laurent expansion of f at w' is

$$\sum_{j,k=-\infty}^{\infty} a_{jk}(w') z_1^j z_2^k,$$

then $a_{jk}(w') = a_{jk}(w)$ by the uniqueness of the Laurent expansion. Thus $a_{jk}(w)$ is locally a constant in R. Since R is connected, we have $a_{jk}(w) = a_{jk}$, a constant which is independence of w. Thus the Laurent expansion of $f(z)$ on R is

$$\sum_{j,k=-\infty}^{\infty} a_{jk} z_1^j z_2^k$$

and it converges uniformly on a neighborhood of any point $z \in R$, and hence converges uniformly on any compact subset in R.

From Theorem 13, we have

Theorem 14 If R is a Reinhardt domain in $\mathbb{C}^2$, R contains the points whose first coordinate is 0, and the points whose second coordinate is 0, that is the points $z = (z_1, z_2)$ where $z_1 = 0$ or $z_2 = 0$, then the holomorphic function $f(z)$ on R has the expansion

$$f(z) = \sum_{j,k \geq 0} a_{jk} z_1^j z_2^k \tag{5.2}$$

and it converges uniformly on any compact subset in R.

Proof By Theorem 13, $f(z)$ has the expansion (5.1). All $a_{jk} = 0$ when $j < 0$ since R contains the points $(0, z_2)$, otherwise (5.1) can not converge uniformly at a neighborhood of $(0, z_2)$. Since R contains the points $(z_1, 0)$, $a_{jk} = 0$ when $k < 0$. It completes the proof of the theorem.

Theorem 15 (Hartogs theorem on Reinhardt domains) If R is a Reinhardt domain in $\mathbb{C}^2$, and R contains some points whose first coordinate or second coordinate is zero, that is, the points $z = (z_1, z_2)$ where $z_1 = 0$ or $z_2 = 0$, then any holomorphic function $f(z)$ on R can be analytically continued to $R' = \{(\rho_1 z_1, \rho_2 z_2) \in \mathbb{C}^2 \mid 0 \leq \rho_1 \leq 1, 0 \leq \rho_2 \leq 1, (z_1, z_2) \in R\}$. In other words, there exist a holomorphic function F on R' such that $F = f$ if $z \in R$.

The proof of the theorem is obvious. Note that $f(z)$ can be expanded as (5.2) on R, (5.2) converges uniformly on a neighborhood of any $z \in R$. If $(\rho_1 z_1, \rho_2 z_2) \in R'$, then

$$\sum_{j,k \geq 0} a_{jk} \rho_1^j \rho_2^k z_1^j z_2^k$$

converges. Hence (5.2) converges uniformly on a neighborhood of $(\rho_1 z_1, \rho_2 z_2)$. Denoting the convergence function by F, it is the required function.

Finally, we give an example of the Hartogs phenomenon.

Example Let $B_r(0,1) = \{z = (z_1, z_2) \in \mathbb{C}^2 \mid r < |z_1|^2 + |z_2|^2 < 1\}$, then $B_r(0,1)$ is a Reinhardt domain, where $0 < r < 1$. If $f(z)$ is holomorphic on $B_r(0,1)$ then f can be analytically continued to the unit ball $B(0,1) = \{z = (z_1, z_2) \in \mathbb{C}^2 \mid |z_1|^2 + |z_2|^2 < 1\}$.

We discussed the Hartogs theorem on Reinhardt domains as above. By the Hartogs theorem in several complex variables, we may ask the following question: what kind of domains in several complex variables do we need to discuss? Roughly speaking, the answer of this question is that the domains without Hartogs phenomenon are interesting for us. These domains are domains of holomorphy. To study the domain of holomorphy was one of the main topics in several complex variables in 20th century. The readers may refer the books of several complex variables, for example, S.G. Krantz[14], R. Narasimhan[15].

In this chapter, we state and prove the results of several complex variables in $\mathbb{C}^2$ only, but it is no any difficult to extend all these results to $\mathbb{C}^n$.

REFERENCES

[1] S. Gong and S.L. Zhang, *Concise Calculus* (University of Science and Technology Press, Heifei, 1997, 3rd Edition) (In Chinese).

[2] L.V. Ahlfors and L. Sario, *Riemann Surface* (Princeton University Press, Princeton, New Jersey, 1960).

[3] H.H. Wu, I.N. Lu and Z.H. Chen, *Introduction to Compact Riemann Surface* (Science Press, Beijing, 1981) (In Chinese).

[4] G. Springer, *Introduction to Riemann Surface* (Addison-Wesley Publishing Co., Massachusetts, 1957).

[5] L.V. Ahlfors, An extension of Schwarz's lemma, *Tran. Amer. Math. Soc.* **43** (1938) 359–364.

[6] R.M. Robinson, A generalization of Picard's and related theorems, *Duke Math. J.* **5** (1939), 118–132.

[7] H. Grauert and H. Reckziegel, Hermiteschen metriken and normale familien holomorpher abbildungen, *Math. Z.* **89** (1956), 108–125.

[8] Z. Kobayashi, *Hyperbolic Manifolds and Holomorphic Mappings* (Marcel Dekker, New York, 1970).

[9] L. Zalcman, A heuristic principle in complex function theory, *Amer. Math. Soc. Monthly* **82** (1975), 813–817.

[10] D. Minde and G. Schober, Another elementary approach to the theorems of Landau, Montel, Picard and Schottky, *Complex Variables*, **2** (1983), 157–164.

[11] S.G. Krantz, *Complex Analysis; The Geometric Viewpoint* (MAA, Washington D.C., 1990).

[12] L.V. Ahlfors, *Conformal Invariants* (McGraw-Hill, New York, 1973).

[13] J.B. Conway, *Function of One Complex Variable* (Springer-Verlag, New York, Berlin Heidelberg, London, 1986).

[14] S.G. Krantz, *Partial Differential Equations and Conformal Analysis* (CRC Press, Boca Raton, Ann Arbor, London, Toyko, 1992).

[15] R. Narasimhan, *Several Complex Variables* (University of Chicago Press, Chicago, 1971).

[16] L.K. Hua, *Introduction to Higher Mathematics*, Vol. II (Science Press, Beijing, 1982) (In Chinese).

[17] G.T. Zhung and N.O. Chang, *Functions of Complex Variable* (Peking university Press, Beijing, 1984) (In Chinese).

[18] J.Y. Yu, *Function of Complex Variable* (Higher Education Publishers, Beijing 1980) (In Chinese).

[19] L.L. Fan and C.G. Wu, *Theory of Functions of Complex Variable* (Shanghai Science and Technology Press, Shanghai, 1987) (In Chinese).

[20] L.V. Ahlfors, *Complex Analysis*, 3rd ed. (McGraw-Hill, New York, 1979).

[21] S.R. Bell, *The Cauchy Transform, Potential Theory and Conformal Mapping* (CRC Press, Boca Raton, Ann Arbor, London, Toyko, 1992).

[22] C.A. Berenstein and R. Gay, *Complex Variables, An Introduction* (Springer-Verleg, New York, Berlin Heidelberg, London, 1991).

[23] H. Cartan, *Théorie Elementaire des Fonctions Analytique d'une on Plusieurs variables complexes* (Hermann, Paris, 1961).

[24] C. Carathéodory, *Conformal Representation* (Cambridge University Press, London, 1958).

[25] C. Carathéodory, *Theory of Functions of a Complex Variable*, Vol. I, II, (Chelsea, New York, 1954).

[26] W.H.J. Fuchs, *The Theory of Function of One Complex* Variable (Van Nostrand, New York, 1967).

[27] R.E. Greene and G.G. Krantz, *Biholomorphic Self-maps of Domains*, Lecture Notes, No.1276, (Springer-Verlag, New York, Berlin Heidelberg, London, 1987), 136–207.

[28] M. Heins, *Complex Function Theory* (Academic Press, New York, 1968).

[29] E. Hille, *Analytic Function Theory* (Ginn and Company, Boston, 1959).

[30] A. Hurwitz and R. Conrant, *Funktionen Theorie*, 4th ed. (Springer-Verlag, New York, Berlin Heidelberg, London, 1964).

[31] C. Jordan, *Cour d'Analyse* (Gauthier-Villars, Paris, 1893).

[32] S.G. Krantz, *Function Theory of Several Complex Variables* 2nd ed. (John Wiley and Sons, New York, 1991).

[33] S. Lang, *Complex Analysis*, 2nd ed. (Springer-Verlag, New York, Berlin Heidelberg, London, 1985).

[34] A. I. Markushevich, *Theory of Functions of a Complex Variable* (Chelsea, New York, 1977).

[35] R. Narasimhan, *Complex Analysis in One Variable* (Birkhäuser, Basel and Boston, 1984).

[36] R. Nevanlinna, *Analytic Functions* (Springer-Verlag, New York, Berlin Heidelberg, London, 1970).

[37] B.P. Palka, *An Introduction to Complex Function Theory* (Springer-Verlag, New York, Berlin Heidelberg, London, 1991).

[38] W. Rudin, *Real and Complex Analysis* (McGraw-Hill, New York, 1966).

[39] S. Saks and A. Zygmund, *Analytic Functions* (Elsevier, Amsterdam, London, New York, 1971).

[40] C.L. Siegel, *Topics in Complex Function Theory* (John Wiley and Sons, New York, 1969).

[41] E.C. Titchmarsh, *The Theory of Functions* (Oxford University Press, London, 1939).